Yogo The Great American Sapphire

Yogo The Great American Sapphire

by

Stephen M. Voynick

MOUNTAIN PRESS PUBLISHING COMPANY
MISSOULA

Ninth Printing, March 2016

Historical photographs courtesy of Delmer L. Brown.

Library of Congress Cataloging-in-Publication Data

Voynick, Stephen M.
Yogo : the great American sapphire.

Rev. ed. of: The great American sapphire. 1985.
Bibliography: p.
Includes index.
1. Sapphire mines and mining—Montana—Yogo Creek Region—History.
2. Yogo Creek Region (Mont.)—History.
I. Voynick, Stephen M. Great american sapphire.
II. Title.
TN997.S24V69 1987 338.7'622384'0978662 87-25683
ISBN 978-0-87842-217-3

PRINTED IN THE UNITED STATES OF AMERICA

P.O. Box 2399 • Missoula, MT 59806 • 406-728-1900
800-234-5308 • info@mtnpress.com
www.mountain-press.com

This book is for Lynda.

Other Books by Stephen M. Voynick

The Making of a Hardrock Miner	Howell-North Books
In Search of Gold	Paladin Press
The Mid-Atlantic Treasure Coast	The Middle Atlantic Press
Leadville: A Miner's Epic	Mountain Press Publishing Co.
Colorado Gold	Mountain Press Publishing Co.
Colorado Rockhounding	Mountain Press Publishing Co.

Contents

. . . the real treasures that have come to us in life are the blue pebbles that the gopher dug from the earth–not the treasures we sought in far places.

—from an editorial
The Great Falls Tribune
January 7, 1930

Preface

To most Americans, the word sapphire has an intriguing foreign flavor conjuring images of crown jewels, sultans, turbanned natives, and the steamy jungles of exotic places like Ceylon and Burma. Many Americans, including myself, were quite surprised to learn recently that the United States had suddenly emerged as a commercial source of what some gem experts consider the world's finest sapphire. Even more surprising was that the entire production came from a little-known Montana mine with the unlikely name of Yogo, a mine which, sixty years earlier, had produced $25 million in fine cut sapphires for the British.

My interest in western mining led me to Yogo where I found not only a mine, but a remarkable geological story backed by a century of rich Montana history. That history, in passing from generation to generation, had evolved into a loose collection of colorful frontier folklore and romanticized legends. More fascinating was the real story that lay hidden in disjointed company files and correspondence, dusty geological reports and decades of yellowed newspaper clippings—even in some of the works of Charles M. Russell. Yogo was far more than a common tale of mineral discovery and exploitation; it was the culmination of a forgotten chapter of American history—the search for precious gemstones.

America's frontier expansion coincided with a great period in gemstone history that included discovery of the Siam and Kashmir sapphires, the South African diamond fields, and the British development of Burma's legendary Mogok Stone Tract. Americans headed west in love with, and obsessed by, gold. But they were also aware of the possibility of—and perhaps even anticipated—the discovery of native precious gemstones. Yet, when sapphires were finally discovered in Montana, the same miners who wrote the book on gold were shown to be profoundly naive in matters related to precious gemstones, thus opening the door to eager British gem merchants.

Unlike that of the great gold strikes, the Yogo sapphire story did not die with the frontier. Although yesterday's claim stakes and sluice boxes are gone, equally exciting chapters in the Yogo story are now being written in corporate board rooms, gem industry trade journals, gemological laboratories and, most importantly, in the display cases of thousands of retail jewelry stores across the United States.

Montana's Yogo sapphire deposit is a true bonanza that economically overshadows many major gold strikes, but sapphires, while far more valuable on a weight-for-weight basis, were unlike gold. Gold required merely digging and selling; sapphires demanded marketing, a lesson that hopeful American sapphire miners would take ninety years to learn. Yogo is an historical treasure, but the story of the Yogo sapphires is really just beginning, for only now are South African diamonds, Colombian emeralds and Burmese rubies being belatedly joined by a native American precious gemstone that is every bit their equal—the Montana sapphire.

Introduction

The Yogo Dike

Three hundred million years ago, the land that is now central Montana lay covered by a warm inland sea. For eons, the waters alternately rose and receded, depositing enormous quantities of sediments to slowly build a seabed thousands of feet thick. Among the many geological strata in this ancient seabed was a massive layer of limestone we know today as the Madison formation. As the seabed continued to build, the Madison limestone became a subterranean formation, covered by newer layers of shales and sandstones.

Many millions of years later, the western United States rose in a great geological uplift, eventually coming to rest thousands of feet above the level of the now distant seas. In central Montana, great crustal stresses caused the ancient seabed to buckle, creating the high, rugged ridges of the Little Belt Mountains.

While all that was happening, probably about fifty million years ago, rocks in the earth's mantle, deep beneath the crust, melted to form molten magmas. Some of those magmas had compositions that properly belonged deep in the earth's interior, compositions foreign to the continental crust. One of those exotic magmas rose into a fracture in the Madison limestone, where it crystallized to form a dike. That igneous fracture filling formed an intrusion only about eight feet wide, on average, but at least five miles long.

The magma that formed this dike had been subjected to the incredible temperatures and pressures of the earth's mantle. As it began to crystallize, still at great depth, atoms of oxygen within the seething, molten mass combined with those of aluminum to form corundum, the mineral form of aluminum oxide. By rare coincidence, the corundum formed tiny, perfectly shaped, transparent crystals quite unlike the usual blue-gray prisms. By an even more unlikely coincidence, virtually every crystal contained a minute amount of iron and titanium that imparted to each a beautiful cornflower blue color. After their birth in the fiery womb of the earth, the tiny blue crystals were carried upward in the stream of surging, bubbling magma high into the crust. Within that narrow fissure in the Madison limestone, at the place to be called Yogo, the magma solidified, fixing the blue crystals in time and place within the subterranean dike.

Fifty million or so years passed as the slow processes of erosion sculpted the rocks of central Montana into the mountains and valleys we know today. In time, perhaps as recently as one million years ago, erosion of the ancient seabed exposed the long-hidden Madison limestone and the dike with its precious load of sapphires. The elements deteriorated the dike rock very quickly into a crumbly earth and at some moment the first of the billions of tiny, blue crystals tumbled free from its matrix. Man, who had yet to appear on the North American continent, would call this splendid creation of nature sapphire.

With infinite patience, nature continued to sculpt the land to her liking. From the Little Belt Mountains, a stream rushed east transecting the dike and cutting deeply into the sapphire-bearing rock. Increasing numbers of the blue crystals became part of the stream gravel. As they tumbled over and over in the water and gravel, they became rounded and scoured into dull, translucent blue pebbles. The blue pebbles, being somewhat heavier than the common gravels, worked their way to the bottom of the gravel bed, where they became concentrated with another heavy mineral, bits and flakes of yellow placer gold.

When the Crow and Sioux Indians, the wandering hunters of the high plains, came to central Montana, they found herds of bison and antelope on the grassy plains, beaver in the streams, and deer, elk and bear in the dense pine forests that cloaked the Little Belt Mountains. Hunting parties camped often along the creek that transected the ancient dike, but the hunters sought only meat and hides, not the gold and blue pebbles that waited in the gravels of Yogo Creek. But the time was already at hand for the coming of the white man to central Montana. Although they came searching for gold, they would find the blue pebbles. The Yogo sapphire was ready for its role in history.

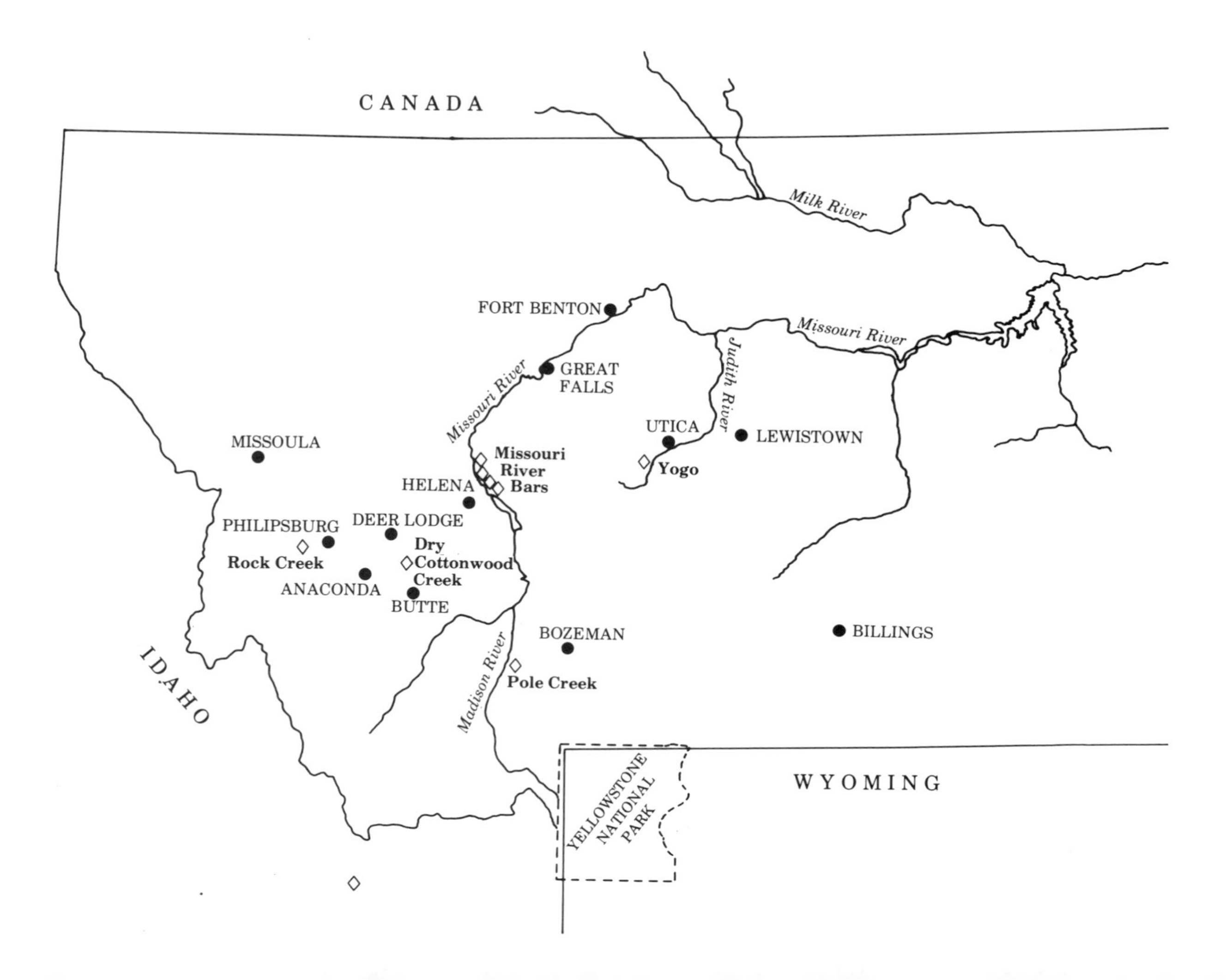
CANADA
Milk River
FORT BENTON
Missouri River
GREAT FALLS
Missouri River
Judith River
UTICA
LEWISTOWN
Yogo
MISSOULA
Missouri River Bars
HELENA
PHILIPSBURG
DEER LODGE
Dry Cottonwood Creek
Rock Creek
ANACONDA
BUTTE
BOZEMAN
BILLINGS
Madison River
Pole Creek
IDAHO
YELLOWSTONE NATIONAL PARK
WYOMING

Chapter 1

The Blue Pebbles

One of the least known aspects of the great frontier hunt for mineral wealth that opened the western United States was the search for precious gemstones. In 1541, the Spanish explorer Francisco Coronado trekked north from Mexico hoping to repeat the earlier successes of Cortes and Pizarro in acquiring huge quantities of gold, turquoise, amethyst and emeralds. According to fabricated Indian tales, his goals, the mythical cities of Quivera and Cibola, contained vast treasures of gold and gemstones. Coronado was the first to fail in the search for native American precious gemstones and, in the centuries to come, he would be joined by many others. When English settlers reached Virginia in the early 1600s, they had been instructed "to searche for gold and such jeweles as ye may find." Although the English had no more success than did Coronado, an enduring pattern had been established: the search for precious gemstones in America would be firmly tied to, and often overshadowed by, the search for gold.

Important gold strikes were made first in North Carolina in 1799, then in Georgia in 1829. The southern Appalachian prospectors brought America its first experience with native precious gemstones. By 1840, diamonds, rubies and sapphires had been occasionally found in the sluice boxes of North Carolina gold miners. Although the diamonds were quite small, a few were cut into attractive and valuable gems. There were not enough diamonds to be of economic significance, but their presence was both tantalizing and encouraging to prospectors.

Sapphires and rubies, however, were another matter. Both were gem forms of the common mineral corundum, or aluminum oxide, which was first identified in the United States in 1819. In the early 1840s, corundum, already recognized as a valuable industrial abrasive, was discovered in quantity in North Carolina. When the bonded grinding wheel was developed in 1860, demand for industrial corundum grew. In 1871, the nation's first commercial corundum mine was opened at Corundum Hill, North Carolina. The Corundum Hill mine production stimulated American interest in native precious gemstones for, along with the large, opaque crystals of industrial-grade corundum, came many small crystals of ruby and vari-colored sapphire which could be cut into beautiful gems. The North Carolina rubies and sapphires were of little economic significance, but reinforced the belief that a major discovery of precious gemstones in the United States was only a matter of time.

When the forty-niners rushed to California seeking gold in the gravels of the Mother Lode, they, too, occasionally found diamonds in their sluices. Again, the diamonds were small and recovered only as a co-product of placer gold mining, but enough to keep gemstone discovery hopes alive. By 1860, Americans were commercially mining turquoise in the Southwest and even exporting California precious opal to Europe.

The last decades of the nineteenth century were exciting times in the world of precious gemstones. Diamond fields of Brazil were still producing, as were the rich emerald mines of Colombia. A new sapphire deposit was discovered in northern India, and the British were about to bring the traditional Ceylonese and Burmese ruby and sapphire mines to record production. The greatest event, of course, was the discovery of the South African diamond fields. More precious gemstones were now being discovered and mined than in all of history. It was only natural to expect a major gemstone discovery to be made in the American West.

Most American prospectors, however, were totally gold-oriented; practically speaking, they knew nothing of the occurrence and mining of precious gemstones. The grand appeal of gold rested in the utter simplicity of prospecting, identification and mining. The distinctly colored, very malleable and extremely heavy metal could be positively identified by the simplest means. Mining was a matter of simple gravitational separation by washing with water. And any prospector with a practiced eye could accurately judge the monetary value of a nugget or a pouch of raw placer gold.

But gemstones had none of gold's simplicity. Color, shape and weight of the rough gemstones often only hinted at identity. And determining value was even more of a problem. There were no experienced gemstone miners in the United States and, before the 1880s, there was little authoritative writing on the properties and occurrences of native gemstones. In investigating a suspected gemstone deposit, the American prospector was left to his own devices and imagination. More often, imagination prevailed. The dozens of "Ruby Creeks" and "Ruby Mountains" today are testimony to a garnet discovery that was first thought to be ruby. The many colors of common mineral crystals could pass for anything a hopeful prospector wished them to be, and there were few people to scientifically argue otherwise. Pieces of clear quartz, tumbled into intriguing octahedral shapes, were often erroneously identified as diamond.

The general ignorance of matters gemological was found not only among unschooled prospectors and miners, but even in professional and scientific circles. It was a rare mineralogist or geologist who took more than a casual interest in gemology. Nearly all jewelers were simply gem merchants, yet some passed themselves off as "gemologists" to enhance their professional stature. Many mineral crystals suspected of being gemstones were sent to Tiffany & Co. in New York City. As the nation's most prestigious jeweler, Tiffany & Co. quickly became a recognized authority in the scientific identification and evaluation of native precious gemstones.

The man most responsible for compiling data on native gemstone occurrences and making the basics of gemology available to the American public was George Frederick Kunz. Kunz was attracted to the beauty of minerals as a boy; by age fourteen, he had amassed a notable collection of minerals native to New York City and adjacent northern New Jersey. As a budding mineralogist, Kunz was largely self-taught. He attended, but did not graduate from, New York City's Cooper Union. In 1878, at the age of twenty-two, he submitted his first mineral report to the United States Geological Survey. Four years later, he accepted a position as a gemologist with Tiffany & Co. and began writing regular reports for the Geological Survey. In later years, he would personally visit most mineral deposits in the United States and many in foreign nations. His career would be reflected in

531 articles, reports, professional papers and books and he would be honored with numerous academic degrees.

Kunz's first book, *Gems and Precious Stones of North America*, published in 1890, documented the early American experience with gemstones, noting that many gemstones, of gem quality and considerable value, had been overlooked, unknowingly discarded and even destroyed because of primitive testing techniques. At the same time, worthless "gemstone" discoveries had become the basis for grand tales. Among the many amusing incidents that Kunz recorded was the 1883 discovery of the famed "Blue Ridge sapphire," also known as the "Georgia Marvel." The excited finder took the huge blue crystal to two Georgia jewelers who promptly and positively identified it as a gem quality sapphire. Based on its prodigious weight, they appraised its value at $50,000, a veritable fortune in the 1880s. Kunz believed that the identification was based on the "file test," a frequently misleading gem identification method popular at the time. The assumption was that anything an alloyed steel file could scratch was glass or the worthless, softer crystal of a common mineral; anything not scratched was a gemstone. The heavy hands of the jewelry store "gemologists" shattered countless specimens. The "Georgia Marvel" did not shatter, but it wouldn't have mattered if it had. The "sapphire" that had caused such excitement in Georgia was later found to be a piece of rolled blue bottle glass.

A North Carolina prospector discovered a fist-sized, nine-ounce "emerald" containing many tiny, glittering, internal "diamonds." When a local "gem expert" confirmed this identification, the ecstatic owner refused a quick offer of $1,000. At considerable personal expense, he took his treasure to New York City—only to have the "emerald" identified as greenish quartz riddled with numerous liquid inclusions. His final offer was $5.

Another case was Alabama's "Watempka Ruby," an enormous six-ounce ruby announced as the largest gem ruby ever found in the world. After spending several months in the tight security of a bank vault, fragments of the wondrous ruby were finally sent to New York City for appraisal. Once again, there was only disappointment. The "Watempka Ruby" was a common garnet, and poor quality at that.

The lack of general gemological knowledge made fraud inevitable. During the 1870s, a number of diamond mine frauds were perpetrated in the mining West, the most successful in southwestern

Wyoming. It began in the summer of 1871 when two strangers appeared at the Bank of California in San Francisco. They identified themselves as miners just back from a prospecting trip who wished to deposit in the bank's vault a sealed canvas bag, the contents of which, they hesitantly revealed, were of the "greatest value." Later, bank president William C. Ralston persuaded the two men, Phillip Arnold and Jack Slack, to disclose the contents of the mysterious bag. Before a small group of Ralston's selected wealthy friends, Arnold and Slack dramatically emptied the bag onto a desk top. The group stared in awe at what certainly appeared to be a heap of huge uncut diamonds and rubies.

The silver fortune pouring from the Comstock Mines in nearby Nevada had raised mining speculation in San Francisco to a peak. Ralston's group knew that the mine which produced those diamonds and rubies would bring untold wealth. At Ralston's suggestion, the stones were submitted to a leading San Francisco jeweler who promptly pronounced them genuine. But before committing any capital, the cautious Ralston insisted that two of the group personally inspect the mine site. The pair, which included a prominent local attorney, traveled by train to southwestern Wyoming, where Phillip Arnold then guided them many miles into the desolate Red Desert country. During the last hours of the journey, the guests were blindfolded. When the blindfolds were finally removed, the gem field exceeded even the wildest imagination. The potential investors scrambled about picking diamonds and rubies right off the ground. Upon their return to San Francisco, their breathless report resulted in the immediate establishment of the San Francisco and New York Mining and Commercial Company. The company's first move was to engage a geologist to thoroughly survey the property. His eagerly awaited report stated that only twenty-five men could wash out over one million dollars in diamonds and rubies every month. Next, a group of selected potential investors were permitted to visit the site where, in a mere week, they collected 7,000 carats of rubies and 1,000 carats of diamonds. By this time, Arnold and Slack were unable to resist the pressure to sell out their corporate interests. They did so for nearly $750,000.

The first competent geologist to visit the site was Clarence King who, in 1872, passed through the region while on a government survey of the 40th parallel. King noted, doubtlessly with a smile, that each hole containing a gemstone had been created with a sharp instrument. The gemstones had simply been pressed into the mock crevices in an obvious salting scheme. Finally, the truth came out. The diamonds were "niggerheads"—South African industrial-grade stones of little value. The rubies were garnets, and the prize diamond of 108 carats was quartz. And Arnold and Slack, not surprisingly, had disappeared.

Phillip Arnold was later traced to his former Kentucky home where, under threat of criminal prosecution, he returned $150,000 of the investors' money. But Jack Slack was never found. Apparently he made the first fortune in native American precious gemstones. The entire episode went down in history under the tarnished name of "The Great Diamond Hoax."

By 1890, inspired by the search for gold, prospectors had visited nearly every creek and mountain in the American West. Great strikes of gold and silver had been made, but the gemstone find that would rival the emerald mines of Colombia or the diamond fields of South Africa never materialized. That year, George Frederick Kunz realistically summed up the importance of the American experience with precious gemstones to date.

> Nearly all the known varieties of precious stones are found in the United States, but there is little systematic search for them, as the indications seldom justify the investment of much capital in the search. The daily yield from the coal and iron mines would exceed in value all the precious stones found in the United States during a year.

Although the expected discovery of a major gemstone deposit had not yet been made, prospectors had long been moving closer to it. In 1862, the search for gold was about to open Montana Territory. Idaho miners, picking, shoveling and panning their way into western Montana, finally struck pay dirt at Bannack. Territorial exploration and settlement rapidly followed a string of subsequent strikes that led to Alder Gulch, Virginia City and, finally, in 1864, to Helena. In only three years, gold had swelled the white population of Montana Territory from a few hundred to more than 30,000.

From Helena, the prospectors continued east, soon striking placer gold in the massive gravel benches that lined the Missouri River. With gold found along a fifteen-mile section of the river, the ancient, glacially-formed bars were quickly crisscrossed with the sluices, flumes and trenches of the placer miners and adorned with the white man's names—Eldorado Bar, Spokane Bar, French Bar, Ruby Bar and Metropolitan Bar. Along with the bits and flakes of yellow placer gold, the miners began noticing small, variously colored, translucent pebbles in their sluice concentrates. Since the pebbles were retained in the sluice box riffles, they were obviously heavier than the common gravels. Although all were water worn, many still exhibited a pronounced crystalline structure. There seems to have been little doubt that the pebbles were sapphires. Kunz recorded the date of discovery of sapphires in Montana as May 5, 1865 and attributed the discovery to Ed Collins, "an earnest and reliable prospector." Not only did Collins correctly identify the stones, he also judged their value to be significant, for he was soon shipping them to Tiffany & Co. and M. Fox & Co., both in New York City, for cutting into gems. To investigate the marketability of the Montana sapphires in Europe, the world's leading gem market, Collins even shipped stones to an Amsterdam diamond cutter.

While earlier "discoveries" of erroneously identified gemstones in the East had created considerable excitement, the discovery of the Missouri River sapphires—*bona fide* gemstones—elicited an oddly subdued response. There was neither wild cheering nor a frenzied rush of would-be sapphire miners. If anything, the miners seemed only mildly curious. The first published mention of the existence of the Missouri River sapphires was not even made until 1873 when Dr. J. Lawrence Smith, an authority on sources of industrial corundum, wrote in the *American Journal of Science*:

> About a year ago, a quantity of rolled pebbles were sent to me from the territory of Montana, which upon examination I found to consist primarily of corundum: they were like the rolled pebbles from the ruby localities of the East Indies, each one being a little crystal in itself, more or less abraded on the angles, and being of a compact uniform structure. They were flattened hexagonal prisms with worn edges. They were either colorless or green, varying in shade from a light to a dark green; some were bluish-green, but none red; there were some red pebbles, but on examination they proved to be spinel.[1]
>
> These pebbles are found on the Missouri River near its source, about one hundred and sixty miles above Benton; they are obtained from bars on the river, of which there are four or five within a few miles of each

[1] Smith's "spinel" was actually garnet.

other. In the mining region of this territory on these bars considerable gold is found, it having been brought down the river and lodged there, and the bars are now being worked for gold. The corundum is found scattered through the gravel (which is about five feet), and upon the bed-rock in the gulches from forty to sixty feet below the surface, but it is very rare in such localities. It is most abundant upon the Eldorado Bar situated upon the Missouri River about sixteen miles from Helena; one man could collect from this bar about one to two pounds per day.

I have had some of the stones cut, and among them one very perfect stone of three and a half carats, and of a good green color, almost equal to the best oriental emerald.[2]

My opinion is that this locality is a far more reliable source for the gem variety of corundum than any other in the United States I have yet examined.

Smith, who had visited every other important corundum source in the nation, was correct in his assessment of the Missouri River sapphires, but commercial sapphire mining would not yet become a reality. Many of the sapphires were of industrial rather than gem grade. The market for industrial corundum was not yet able to absorb any new supply beyond that produced by the eastern mines. An even bigger drawback to commercial mining was that virtually all the Missouri River sapphires were "fancies," that is, pink, green, clear or yellow. The most valuable gem sapphires were the blues, of which the Missouri River bars produced very few. Another factor against mining was the territory and the nature of its miners. Montana was still quite isolated from the mainstream of western development, much less eastern and European gem markets. The miners concentrated on what they knew and felt most comfortable with—gold. Even raw gold was inherently valuable; its value did not depend upon mysterious appraisals of gemologists, mineralogists or jewelers, nor upon the intricacies of cutting and polishing. Some Missouri River bar miners took the time to sort out any sapphires; others merely discarded everything in their sluice concentrates that was not gold. The recovered rough sapphires were traded locally at arbitrary, usually low values based upon casual appraisals of color, size and clarity. As a final discouragement, with the notable exception of the Great Diamond Hoax, American capitalists were reluctant to risk much money in any gemstone venture, perhaps in realization that the British controlled the world precious gem markets from mining to market-

[2] "Oriental emerald" was a common term for green sapphire.

ing. For all these reasons, the discovery of the Missouri River bar sapphires was not a major mining event.

Yet many of the Missouri River sapphires were quite beautiful. In *Gems and Precious Stones of North America*, Kunz noted:

> The Montana specimens rarely exceed ¼ inch to ½ inch in length. They are brilliant but usually of pale tints. Two gems are in the Amherst College Collection, which weigh about ⅛ carat each. One is a true ruby-red, and the other is a sapphire-blue, colors rarely met with here. The gems are usually of a light-green, greenish-blue, light-blue, bluish-red, light-red and red, and the intermediate shades. They are usually dichroitic, and often blue in one direction and red in another, or when viewed through the length of the crystal, and frequently all the colors mentioned will assume a red or reddish tinge by artificial light. A fine one of nine carats was found of a rich steel blue. A very beautiful piece of jewelry, in the form of a crescent, was made of these stones by Tiffany & Co. in 1883; at one end the stones were red, shaded to bluish-red in the center, and blue at the other end; by artificial light, the color of all turned red. Perfect gems of four to six carats are frequently met with. . . . Many are found that are never cut, for it involves greater skill, involving much higher cost, to cut sapphire, than gems which are less hard.

In those final lines, Kunz pointed out another reason for a lack of American interest in Missouri River sapphires—no real domestic cutting facilities. The full beauty and value of a gemstone emerges only after proper cutting and polishing. Although a limited number of gemstones could be cut in New York City, American cutters were decades behind the state of the art as practiced by leading European gem cutters. Even if Montana sapphires were mined, it appeared that the real profit would be made by those who bought rough stones, then cut and marketed the finished gems.

Large quantities of rough sapphires were recovered by the Missouri River bar placer miners, but their cut value never exceeded $2,000 per year. Based on an average value of $10 per cut carat, the annual production of cut sapphires was not more than 200 carats, making them of little economic significance. Still, the Missouri River bars were the most encouraging source of precious gemstones yet discovered in the United States. They also first linked sapphires with Montana, an enduring association of names.

* * *

Change was coming rapidly to the West. In 1876, Custer's Seventh Cavalry was annihilated at the battle of the Little Big Horn in south-central Montana. Now, increased Army attention to containing or eliminating the remaining Indians aided the march of white civilization in remote territories like Montana. Western mining was also in transition. The last great western placer gold rush took place in the Black Hills of Dakota Territory in 1876 and the glory days of placer mining, in which an individual with a gold pan and a little luck might make a fortune, were about over. Virtually every gold-bearing stream in Montana had been mined with varying degrees of thoroughness. One of the very few that had been bypassed was located almost at the geographical center of Montana, about one hundred miles east of Helena. This forgotten stream flowed east out of the Little Belt Mountains to empty into the Judith River. The slowly growing number of Judith River Basin ranchers and farmers called it Yogo Creek.

The origin of the name *Yogo* had evidently already become lost. Most agreed it had been derived from an Indian word. Some said it meant "blue sky," others believed it to be a derogatory word for an unmarried, pregnant woman. "Going over the hill" was offered as another possible meaning. Whatever the true meaning of the name, Yogo was of no consequence until the fall of 1878 when placer gold was discovered in the upper creek.

Apparently, this was not the first time gold was discovered in Yogo Creek. According to the September 1, 1879, *Rocky Mountain Husbandman*, then published at Diamond City, gold had been found in 1866 by prospectors who had probably pushed east from Helena.

> Upon a blazed tree in the Judith (Yogo) Mines is registered the name of a party of prospectors who discovered gold there thirteen years ago and being set afoot by the Indians, did not stay to test the mines. The inscription closes with, "We leave here afoot." The names were engraved in wood, then traced with a pencil, and have survived the storms of the last decade well, being perfectly legible. The name of Joe Summy, a gentleman well known in Meagher county, heads the list.

While the presence of hostile Indians may have cut short any 1866 mining venture, the gold strike of 1878 was free to follow its normal course. By spring, 1879, over 1,000 prospectors and adventurers had converged on this latest hopeful El Dorado. Almost overnight, a rough boom camp sprang up to occupy the once pristine mountain gulch. Depending upon whom one spoke with, its name was either Belt City, Yogo City or Hoover City. The official name soon became Yogo City at the insistence of the U.S. Post Office Department.

Yogo City was destined to be one of Montana's most poorly documented gold camps. Thanks to the *Rocky Mountain Husbandman*, some description of the early camp is recorded in this article which appeared on November 20, 1879.

> The new mines are settling down to a steady thing, a good camp being assured. There were fifty-one cabins at this time, two saloons, one store and one boarding house. There were 200 men in the camp, 150 of them would winter there. Many of these had laid in a winter's supply of provisions.

A far more glowing report of Yogo City appeared in the *Husbandman* on December 11.

> The news from the mines is flattering in the extreme. It is estimated that there are now fully 100 houses up and many more building. The houses are of a very substantial character and indicate that the builders had implicit faith in the mines. Everyone seemed to be at work in earnest. Specimens showing free gold were numerous, and the quartz prospectors were as sanguine as the placer miners. Specimens of black galena assaying 86 ounces to the ton were also on exhibition.
>
> Labor was worth from two to three dollars per day; lumber 100 dollars per thousand feet, flour $8 per sack. The new town was building up rapidly and was being put in good shape. This, in addition to what we have previously published, pretty clearly establishes the belief that Yogo will prove a second Confederate Gulch or Alder Gulch, and will at an early day, be the home of a thousand or more prosperous miners.

But Yogo City had little else going for it than optimistic newspaper reports; within three years it was a ghost town of rotting cabins with a population of less than fifty. The *Husbandman*, like all newspapers serving the western mining regions, felt it a civic duty to embellish its mining reports, a journalistic excess justified by the necessity to attract outside investment capital. Any discouraging reports, however true, were rarely printed. About 1,000 people had rushed to Yogo but, finding only mediocre gravels, left just as quickly in search of better "dirt" elsewhere. By 1881, even the United States Post Office had closed shop. All that remained in Yogo City was a handful of determined hardrock miners driving narrow tunnels and sinking shallow shafts in pursuit of the elusive mineral veins.

The greatest mineral strike of the Yogo mining district had already been made, but went unrecognized. During 1879, the first full year of the brief boom, prospectors had searched every inch of Yogo Creek and its tributaries. Their search included the lower section of the creek, about three miles downstream from Yogo City and just below the point where a strange geological formation cut deeply through the limestone cliffs to transect the creek. A lack of water hindered their sluicing, but in the little gold they found, they also noted the presence of numerous tiny, translucent blue pebbles. If any of those nameless miners had come from the Helena region, or perhaps the Missouri River bars, they would have noticed a similarity between the blue pebbles of Yogo and the Missouri River sapphires. The obvious difference would have been the color; instead of the vari-colored stones from the Missouri River, nearly all Yogo pebbles were a uniform rich blue. Very possibly, the blue pebbles were recognized as sapphires. But the miners had no way of knowing that the color of these sapphires would match that of the best blue oriental sapphires; they knew only that the Missouri River sapphires had made no one a fortune. The miners were concerned only with gold, and since lower Yogo Creek had little of the yellow metal, they tossed the blue pebbles back into the gravels and departed.

If Yogo City had little gold, it had more that its share of colorful characters. One 1879 arrival noticed that access to the camp was limited to a single winding road coming in from the north. A much easier trail came in from the cow town of Utica, eighteen miles to the east, but terminated abruptly three miles below Yogo City at the base of Yogo Gulch. The obstacle was a narrow cleft in a limestone cliff only a few feet wide. Dudley Hawkins, seeing a better way to make money than shoveling gravel from morning until night, blasted away the limestone to make room for the passage of a wagon. Feeling such civic improvement warranted remuneration, Hawkins erected a moveable tollgate—a trimmed pine trunk—across the new road. His symbol of authority as a toll collector was a Colt revolver. Collecting tolls of fifty cents per horse and rider and one dollar per wagon, Hawkins did much better than most Yogo City miners. The high ground above the tollgate became known as Tollgate Hill and the Tollgate cliffs became a local landmark.

The tollgate at the base of Yogo Gulch. The natural, narrow cleft in the limestone was widened by blasting in 1879 to allow for passage of wagons on the road from Utica to Yogo City. A toll of fifty cents per horse and rider and one dollar per wagon was charged. Today, the road to the American Mine passes through Tollgate.

The most familiar figure in Yogo City was Millie Ringgold, more popularly known as "Nigger Millie" or "Black Millie." Born a slave in Maryland, Millie Ringgold took her emancipation after the Civil War and journeyed up the Missouri River as the nurse and servant for a United States Army general. When the general was transferred back East, Millie elected to remain in Montana at Fort Benton. When Millie heard of the Yogo gold strike, she bought a wagon and two condemned army mules, loaded up with provisions and a barrel of whiskey, and headed for the Little Belt Mountains. By September, 1879, she had established a restaurant, saloon and a small hotel at Yogo City. During the brief boom, Millie profited handsomely, then began buying claims as the disappointed miners left. Much was made of her makeshift musical talents. The newspapers reported that she could "make music on anything, even two sticks or a washboard if those were the only instruments about." The miners remembered her as short and heavy, with shiny black hair and a double row of front teeth. The latter was a source of wonderment to the miners who, it

"Aunt" Millie Ringgold, the ex-slave who joined the gold rush to Yogo in 1879 and who lived in old Yogo City for the last twenty-seven years of her life. The structure in the background is a waterwheel that was used to power a crusher at the Weatherwax Mine at old Yogo City.

was said, constantly begged her to open her mouth so they might admire that rare dental phenomenon. After the bust, Millie continued to tend to her declining businesses and work her claims, never losing faith in the Yogo mines. Millie Ringgold, a long way from her roots as a Maryland slave, would spend the rest of her life in the decaying, forgotten gold camp of Yogo City.

Another notable resident of Yogo City was Jake Hoover, a hunter and prospector who would become an indelible part of Yogo lore. Jacob Hoover was born in Eddyville, Iowa, in 1849. Uninspired by the depressing Iowa farm life, Hoover left home at the age of sixteen and headed up the Missouri River for the booming gold camps of western Montana Territory. For ten years, Hoover drifted around western Montana, prospecting and trying various odd jobs, including ranching. He is believed to have made one or two substantial gold stikes, for he became known about the gold fields of Deer Lodge, Gold Creek and Helena as the "Lucky Boy." If Jake Hoover had a nose for gold, he also had a notorious inability to profit from his discoveries, a shortcoming that would plague him all his life. Although he never found his fortune in gold, he learned frontier skills and traits very well. By the time Jake Hoover was twenty-one, he could ride, shoot, fight, hunt, trap and drink with the best men in Montana. Fiercely independent, Jake was suited to work only for himself. By 1878, he had drifted into the Judith River Basin, liked what he saw, and settled down on a small homestead in Pig-Eye Basin near the South Fork of the Judith River, not far from Utica. Putting his skills to good use, Jake began hunting and trapping in the Little Belt Mountains, selling meat to local farmers and ranchers and delivering hides, furs and antlers to the T. C. Power & Brother trading post at Fort Benton.

When gold was discovered in nearby Yogo Creek in 1878, Jake Hoover didn't have far to go. Although some historians have attributed the Yogo discovery to Jake, it appears that he found no gold whatever. Instead, Jake Hoover kept busy supplying the miners with fresh meat, acting as the district's first recorder and doubtlessly leading the move to name the booming little camp Hoover City. Jake only spent a season at Yogo City, moving back to his cabin in Pig-Eye Basin to resume his hunting and trapping, and to begin establishing a small ranch. But Yogo Creek was not through with Jake Hoover just yet. Many years would pass before Jake would return to Yogo to make the biggest discovery of his life. Although he would still be searching for gold, the color of his discovery would be blue.

As Yogo City slid slowly into oblivion, camps on the Missouri River bars were still producing both placer gold and, for the miners who wished to keep them, sapphires. In 1883, a Helena miner displayed a large, clear stone he found on the bars, uncertain whether it was a sapphire or a diamond. The *Helena Herald* saw the stone as a sign that commercial gemstone mining would soon begin.

> It is not improbable that the accidental discovery of this gem will lead to an industrious search among the tailings and placers of every district in Montana, and it may not be long before diamond-washing will become one of the important industries of the state.

The *Herald's* optimism over the commercial potential of the Missouri River sapphires was still a bit premature, for the focus of Montana's mining industry remained fixed on metals. While Montana sapphires attracted little interest at home, they were becoming a subject of considerable discussion 5,500 miles away in London, England. In 1890, Edwin W. Streeter and Horatio Stewart, two leading London gem merchants, journeyed to Montana to quietly inspect the Missouri River bar placer workings. The following year, many Montanans were surprised to learn that 3,900 acres—over six square miles—including the prime sapphire-bearing placers on Eldorado Bar, had been acquired by British capitalists. The interest in commercial sapphire mining had actually begun in 1888 in Montana. F. D. Spratt and his brother, A. N. Spratt, of York, a small camp northeast of Helena, suspected that the time for sapphires was nearing. Accordingly, they bought up the available, well-worked river bar claims with a hope to profit by mining both gold and sapphires. Needing additional capital, as well as gem marketing expertise, the Spratts solicited the participation of some London gem merchants. By 1891, the Spratts were out and the British were in.

Montanans had no grand illusions about the marketability of the Missouri River vari-colored sapphires, being clearly aware that they were not the classic blue colors demanded by European markets. But the *Helena Herald* quickly pointed out how this would be overcome.

> In order to make the business of mining these Montana sapphires profitable, the public taste must be educated to appreciate their un-

> usual colors, and this the English company will set about doing. Very beautiful effects in jewelry can be produced by combining the gems of different colors in rings, bracelets or necklaces. It is reported that a necklace of Montana gems was recently worn by a duchess at a ball given by the Prince of Wales and that it attracted much notice from connoisseurs.

The British-owned company formed to mine the sapphires was the Sapphire and Ruby Mining Company of Montana. While Montanans knew nothing of what transpired in London, they could see visible progress on the bars. The Sapphire and Ruby Mining Company purchased an additional 140 acres opposite Canyon Ferry, where it was announced that a powerful steam pump would be installed to raise river water to the bars for hydraulicking and sluicing operations. Thousands of feet of iron pipe and other construction materials began arriving. On paper, the British plan appeared well thought out: enough gold would be mined to cover operational costs, leaving the sapphires as clear profit. The eagerly awaited prospectus of the Sapphire and Ruby Mining Company of Montana, finally issued in October, 1891, attracted considerable attention in Montana as well as the East where much of the capital would be raised. In the prospectus, Edwin Streeter, the prominent London jeweler who was retained as a consultant to the company, stated:

> Excepting only the South African Diamond Fields, I consider the Sapphire and Ruby Mines of Montana to be the most important gem discovery of modern times.

The authorized capitalization of the company was an impressive $2 million; the list of original investors included a dazzling array of dukes, marquesses, earls, viscounts and distinguished British businessmen. As intended, the prospectus ignited sapphire fever and the investment dollars began rolling in, with each contributor believing fully that the time had come for American sapphires to take their rightful place among the world's great precious gemstones.

The close attention given to the Sapphire and Ruby Mining Company by Montana newspapers and trade journals revived a few old frontier mining rivalries. The first punch was thrown by the Denver-based *Mining Industry and Tradesman*, which took a condescending view of the British mining plans.

> The truth about the Sapphire and Ruby Mining Company is that the gravel does carry sapphires as has been known in Denver for years. But jewelers here know they are so badly off color that they scarcely have any value as gems. They are greenish, greenish brown, greenish yellow and pure white. . . . Our English friends are welcome to their purchase. Had it been very desirable, it would have been owned by Colorado men long ago.

The *Montana Mining Review* immediately rose to the defense of the sapphires, pointing out that for twenty-eight years many superior stones had been found which were now "in scientific cabinets or in the dressing cases of fair ladies and princesses of the fashionable world." After recognizing that the opinion of London gem experts was somewhat more reliable than those of a few Denver jewelers, the defense went on:

> While there is nothing to be gained in overestimating the richness of our mines, much may be lost in disparaging their real merits. If our Colorado friends have failed to get an interest in these sapphire mines, so disparagingly mentioned by our contemporary, we can assure them that we have some left far better even than the description given of those sold to our English neighbors.

The Sapphire and Ruby Mining Company stayed in the public eye, acquiring 1,500 more acres with water rights which now reportedly gave the company gem mining rights along fifteen miles of the Missouri River. At the first well-publicized shareholders' meeting in March, 1893, the directors displayed a spectacular collection of sapphires and rubies. One glass case contained over 1,000 carats in uncut gemstones, including a twenty-carat sapphire and several prized pigeon's blood rubies, one weighing a remarkable nine carats. Edwin Streeter displayed a box of earth with both gold and sapphires visible, which he claimed to have personally dug from Eldorado Bar. Streeter then announced to the smiling investors, "One fact is now most thoroughly established. That the Missouri River bars near Helena, known as Eldorado, Spokane, Ruby and French, contain gems of great intrinsic value and unparalleled beauty, true symbols of Montana's position among the Mountain States." When it came to financial business, the directors explained that over 300,000 carats of gemstones had been shipped to New York and London during the past year. There had been—ahem—no cash return yet, of course, for the stones were still in the cutting stages.

That final statement prompted a small group of shareholders to investigate. By January, 1894, they had uncovered fraud. Only one-tenth of the $2 million capitalization had actually been raised, and total development expenditure was only $20,000. The highly-publicized "eight-mile-long water diversion tunnel" was really an unfinished five-foot-wide ditch. The list of distinguished British investors had never purchased ordinary investment shares—they had been given a small honorarium to accept "founder's shares" in return for the use of their names. Edwin Streeter, while humiliated, was not guilty of criminal wrongdoing; he had been baited with quality oriental sapphires and rubies. As a jeweler, and not a gemologist or geologist, his deception regarding the origin of the oriental gemstones had been a simple matter.

In the end, investors lost about $200,000 and the Spratt Brothers had a drawer full of worthless stock certificates. Another loser in the fraud was the fragile, neophyte image of Montana sapphires. The British, after the fraud was exposed, were not held in any great esteem in the Montana mining camps. Still, in future years, the British would continue to lead the way in recognizing the commercial potential of Montana sapphires and in committing risk capital to another sapphire enterprise that Americans would hesitate to support.

* * *

Although the Missouri River bars hosted most of Montana's early sapphire excitement, encouraging discoveries were also made elsewhere in the state. Placer miners searching for gold found sapphires on Dry Cottonwood Creek, twelve miles northwest of Butte, in 1889. The deposit was highly concentrated, for some gold pans yielded thirty stones. In 1893, "about twenty-five pounds" of sapphires were found during a single week's work on the broad, 2,500-acre placer deposit. The predominate colors were pale green-yellows and various shades of aquamarine.

Sapphires were next discovered in 1892 at Rock Creek, sixteen miles southwest of Philipsburg. In 1894, United States Geological Survey investigators reported the sapphires as "exceedingly plentiful," sometimes as high as sixty stones to the gold pan. Rock Creek

Old Yogo City as it appeared in 1918. The camp was founded during the brief boom of 1879 when Jake Hoover was the district recorder. After a brief revival in 1894, Yogo City began its final decline. Little remains of the camp today.

sapphires occurred in a broad color range and George Frederick Kunz considered them to be among the most unusual he had ever seen. Many pale or colorless stones exhibited the asterism of the star sapphire or a silky chatoyancy, leading Kunz to believe they might be a commercial source of star sapphires. Kunz described the Rock Creek sapphires as:

> . . . remarkable for their small colored spots which, when properly cut, change the entire stone to yellow or brown. The red stones are pale, but pronounced rubies, many of them intensely brilliant; the yellows, many tints of brown, blue greens, reds and other colors are distinct from those found at any other locality, and all of the colors are rendered more brilliant by artificial light.

Sapphires were again discovered by gold miners in Quartz Gulch, near Pole Creek, southwest of Bozeman. Pole Creek yielded the largest known specimens of crystalline corundum in Montana, including a 1,120-carat (eight-ounce) specimen showing good red and green colors and a 588-carat piece of ruby corundum. Neither specimen was of gem quality.

By 1894, gold miners had discovered every deposit of gem quality sapphire in Montana, with a single exception. This last, imminent discovery would be the long-awaited bonanza of precious gemstones that had eluded Montana miners for thirty years. It would be made near a forgotten gold camp called Yogo City.

During the late 1880s, Yogo City was alive only in the newspapers. When a branch of the Montana Central Railway was built from Great Falls south into the Little Belt Mountains, many predicted a resurgence in local mining. Many feature articles on the Little Belt Mountain mining camps began appearing in the press. This 1888 *Great Falls Tribune* headline on Yogo City was typical.

THE YOGO CAMP
THE MINES OF THIS NEW ELDORADO
IMMENSE DEPOSITS OF FREE MILLING GOLD ORE
VARYING IN WIDTH FROM FIFTY TO ONE HUNDRED FEET!
A Short Description Of The Mines And What Is Being Done On Them—A Live Place!

The descriptions of the active mines gave a picture of bustling crews of miners taking out bonanza-grade ore and expecting any day to strike the lode that would even make headlines in the East. The real Yogo City, however, was much different. The brief boom of 1879 was already a ten-year-old memory. Now the summer population never exceeded twenty-five miners. During the long Montana winters, the population dropped to less than a dozen, among whom old Millie Ringgold was still the community's most prominent citizen. Yogo had been one of the state's least productive mining districts; realistically, there was little reason for anything to change now.

Of all those who passed through Yogo City, few names would endure in the history books. One that would was Jake Hoover's, the

Yogo district's first recorder. Jake's grip on fame was assured not by any gold discovery, but by a close acquaintance. By 1880, Jake Hoover had left Yogo City and returned to his little ranch in Pig-Eye Basin where he resumed his hunting and trapping. One summer evening in 1881, Jake was returning from a hunting trip in the Little Belts. As darkness fell, he drew up his string of pack horses and made camp on the Judith River. Noticing another campfire a short distance away, Jake strolled over to investigate. He found a sixteen-year-old boy in one of the "sparsest" camps he had ever seen. Almost twenty years later, that boy would vividly describe the meeting for the *Great Falls Tribune*.

> All I owned in the world was a brown mare and a pinto pony. I rode the mare and used the pony to pack my bed, which was very light. With no money or grub, life did not seem joyful, and I felt mighty blue, but leaving the stage station (Utica) I rode a short distance up the Judith River and made camp. While I was wondering where my next meal was coming from, a rider with several pack horses appeared and made his camp on the river near mine. I recognized him as Jake Hoover, whom I had seen several times. After getting his packs off he strolled over to my camp and looked it over.... After surveying my camp, Jake asked: "Where do you keep your grub?"
>
> "I ain't got none," I answered.
>
> Then I told him my troubles. He listened until I was through, and while I was talking I couldn't help feeling that he would be my friend.
>
> "Well," he said, "if you want to, you can come with me...".

Jake Hoover first knew his new acquaintance as "Kid." Kid had been born in St. Louis, growing up near the waterfront where the river steamboats began their long journey to the mystical western lands of bison, Indians, shining mountains, cowboys and gold strikes. Very early, Kid developed an incurable case of wanderlust and his schoolwork took second place to frontier dreams. His enrollment at a strict New Jersey military school didn't help. Finally, his parents gave their permission for him to go west. At that time, a friend of his father's, Wallis "Pike" Miller, who owned a Montana sheep ranch, was visiting St. Louis. When Pike returned to Montana, Kid went with him. After a six-week journey by train, stagecoach, wagon and horseback, they arrived in the Judith River Basin—where Kid's romantic frontier dreams came crashing down to earth. Sheepherding was the worst job imaginable. After only a few months, Kid quit.

Dejected, confused and broke, he had ridden aimlessly up the Judith River where, on a Montana summer evening, he chanced to meet Jake Hoover.

Hoover, who also had left his home at age sixteen, may have seen some of his own youth in Kid, for they became immediate friends. Kid took a bunk at the ranch in Pig-Eye Basin and rode by Jake's side for two years, hunting in the summer and fall, tending trap lines in the winter. Hoover taught Kid the basic frontier skills, but his greatest gift to his young partner was a profound respect for the country—the cowboys, Indians, horses, game animals and the land itself. In two years, Kid had become as much a part of the Montana frontier as Jake Hoover.

Jake was always amused by his partner's hobby. In every spare minute, Kid would be drawing or painting. With art supplies in short supply in the Judith River Basin, Kid used pencil, charcoal and the few paints available, often putting his impressions of Montana life on bleached boards in place of canvas. Hoover proved an excellent critic, subtly suggesting correction in the proportions of the horses, bison and deer that would become a prominent part of the young artist's future work. In 1882, Kid sold his first painting—a watercolor on a wooden board that was purchased to decorate the wall of a Utica saloon. A year later, Kid left Pig-Eye Basin to fulfill his dream of riding herd on a nearby ranch. After 1883, Kid frequently returned to Pig-Eye Basin to visit his friend and take advantage of Jake's generosity. Hoover later recalled, "My fondest memories are of my old homestead in the Pigs Eye Basin. . . . In riding the grub line, he (Kid) was a frequent visitor of mine for weeks at a time." During these years, Kid—Charles Marion Russell—continued to develop his artistic skills, moving always closer to recognition as one of the frontier's premier artists.

If Charlie was to become a great artist, Jake Hoover was no less a great hunter. Old-timers in the Judith River Basin agreed that no man could challenge Jake's abilities with a rifle and on the game trails. Charlie Russell would later remember the man who taught him the ways of the Montana frontier.

> As I remember him then, Jake Hoover was a man of medium height, with thick, curly, brown hair, which he wore quite long, a moustache and several months growth of beard. He wore a light, soft hat, blue

The Judith round-up crew in the 1880s. After living for two years at Jake Hoover's ranch at Pig-Eye Basin, Charlie Russell fulfilled a boyhood dream of riding herd. Russell is shown seated on the left nearest the camera.

> flannel shirt, duck pants with boots. His spurs were short-shanked, with broad steel bands. He never used a cartridge belt, but instead, a plain leather strap on which hung a knife scabbard holding two butcher knives. His cartridges were always carried in a pouch, either in his pocket or hanging under his belt. His gun was a .44 Winchester rifle which he packed across the saddle in front of him in a horse sling, but in the game country, he carried it loose in his hands. . . . He could empty a Winchester faster than any other man I ever knew, never taking it away from his shoulder once he started shooting. . . . Jake would see game where there was nothing visible to me, and he was always right about it. . . .

Using first a .44-40 Winchester Model 1873 and, later, an even heavier .45 caliber Winchester which, as a friend noted, "had enough power to put a hole through an elephant," Jake could knock down running elk at distances at which other men wouldn't waste a cartridge. I. David Finch, a Utica resident who, in his youth, knew

Hoover, also recalled Hoover's remarkable abilities in a 1950 interview.

> Hoover could fire a rifle faster than any man I ever saw. . . . When he spotted game he would put a lot of cartridges in his hat and hold the hat in his teeth. On two occasions, at least, I've known of his killing four bears at a time, two old ones and two cubs.

Hoover himself took considerable pride in his reputation. "I was never out of meat," he once recalled. "I have the credit of killing more bears than any other living man in that part of Montana. I know positively, by memorandum I have kept, that I have killed one hundred and eight bears. . . ." Jake was a complex man for, although hundreds of deer, elk and bear fell before his thundering Winchesters, he also demonstrated considerable sensitivity for animals. Charlie Russell once recalled the story of Jake's pet pig, a classic in Montana folklore, which was recorded in Frank Bird Linderman's *Recollections of Charley Russell* (University of Oklahoma Press, 1963).

> Hoover had a way with animals, Russell recounted in his story of Jake's pig: "every livin' thing around here liked old Jake." Hoover got a "little suckin' pig" in a trade for elk meat and soon enough "that little pig was follerin' Jake wherever he went, up hill and down again, just like a dog. He got to be an awful nuisance." One day after the pig had messed up the cabin and Jake and Charlie were low on grub, Hoover insisted they slaughter the pig. When he approached the pig, it rubbed its nose on his leg and Jake barked: "Git away from here, damn ya. Ya think this is a friendly visit?" Hoover turned to Charlie and asked him to dispatch the pig. Charlie replied: "Not by a damn sight, he ain't my pig." Jake pulled out his Winchester, climbed up the hill, and blasted the "rooter" with one shot. "Old Jake came down off the hill, leaned his rifle against the pen, an' stuck the pig. I can see him yet ashakin' the blood off the knife. "I don't reckon he saw me or knowed who done it, do you, Kid?" he asks, lookin' sorry as a woman.

Hoover also demonstrated a similar compassion for his fellow man. Although he was always short of money and in debt to the T. C. Power trading company at Fort Benton, Jake Hoover never turned his back on a man who was down on his luck. Charlie Russell was one of many who found hospitality at Pig-Eye Basin. Equally proud of his generosity as he was of his hunting skills, Jake later commented, "I have befriended and fed many men at my old homestead, in fact, the latch string of my cabin was always at the service of anyone and I have never charged a cent."

Along with Jake Hoover's many admirable qualities, he also had two shortcomings which became legendary—drinking and womanizing. Charlie Russell, for twenty years a noted and accomplished drinker, apparently developed and perfected that dubious skill under the expert tutelage of Jake Hoover. Jake's drinking seems to have been somewhat safer than his womanizing. In one story, one of Jake's women used his own Winchester to shoot through the wall of a cabin where Jake was entertaining another woman. One of the bullets was said to have "burned a streak across the woman's stomach as slick as a branding iron could have done." In another episode, a hired

Breathitt Gray's saloon, one of three saloons in Utica in the 1880s. If the many stories are to be believed, both Jake Hoover and Charlie Russell spent some leisure time in establishments such as this.

sheepherder had eyes for Jake's cook, a woman Jake was living with in his cabin at Pig-Eye Basin in 1886. Jake fired the herder who later returned, bent on revenge. Armed with a rifle, the herder took a position in an old shack overlooking Jake's cabin, patiently waiting for the opportunity to settle the score. To either discourage or eliminate the angry herder, Jake set a fifty-pound bear trap beneath the shack window. By his own account, Jake nearly stepped in "the damn thing" himself. In the end, all he ever caught was one of his ranch dogs.

In 1892, Charlie Russell hung up his spurs and left the Judith Basin ranches to devote full time to his art in Great Falls. His years with Hoover left an enormous impression, as evidenced by many of his later paintings. In the watercolor *When I Was a Kid* (1905), Russell portrayed himself on horseback with Hoover, packing out game along a mountain trail that bypassed the tollgate on the way to Yogo City. In three other watercolors, *A Doubtful Guest* (1896), *The Deerslayer* (1893), and *Hunter's Rest* (1896), Jake and Charlie are pictured together in the hunting camps. In *Christmas Meat* (1915), Hoover is standing in front of his Pig-Eye Basin cabin greeting another hunter who has brought a slain deer for the holiday dinner. One of Russell's most famous works was *A Quiet Day in Utica*, which captured the atmosphere of Utica through the various characters who frequented the little cow town. Russell originally named the painting "Tin Canning a Dog," for it depicted a practical joke of the time, that of tying an empty five-gallon can behind a dog to stampede horses in the street. Jake Hoover is shown laughing and slapping his knees in obvious approval of the joke. Russell himself is smoking a cigarette while leaning against a hitching post. Behind Jake is old Millie Ringgold who, apparently, in Russell's imagination, had journeyed down from Yogo City to purchase supplies.

Just after Charlie Russell had settled in Great Falls to seriously pursue his art career, there was a brief revival in old Yogo City when two small mines struck small lodes of higher-grade ore. The population swelled to 200 in the summer of 1894, enough for the Post Office Department to reopen the Yogo City Post Office. The boom lasted only a few months. By fall, the Post Office had closed again and the ruins of Yogo City belonged only to Millie Ringgold and a dozen remaining miners.

In A Doubtful Guest, *painted in 1896, Charles M. Russell recalls his days in the hunting camps in the Little Belt Mountains. This scene would have taken place about 1880-1882. Jake Hoover, shown at left, would have been about 33 years of age; Russell (standing), would been about eighteen. Russell claimed to have learned most of his frontier skills under the tutorage of Jake Hoover.*
Amon Carter Museum, Fort Worth, Texas

Jake Hoover had not been doing much mining for the past fourteen years, but the flurry of activity at Yogo City seemed to have renewed his interest in gold. Taking his rusted picks and gold pans, Jake set out again to search for the wealth that had eluded him since he first reached the Montana gold fields nearly thirty years before. Jake headed again for Yogo Creek. Knowing that the upper creek, near Yogo City, had been mined out, Jake concentrated on the lower sections of the creek, just downstream from that strange geological formation that cut through the limestone cliffs and transected the creek. The lower creek had been prospected during 1879. A bit of gold was reported to have been found, but the lack of water had prevented any real mining. Jake thought the area was worth a look. It was here, in the autumn of 1894, that Jake Hoover found gold.

A Quiet Day in Utica, *painted in 1907, was one of Charles M. Russell's most famous works. Originally titled* Tin Canning A Dog, *the painting depicts a practical joke of the time, that of tying an empty five-gallon can behind a dog to stampede horses in the street. The painting shows Russell himself and several characters leaning on a hitching rail smoking a cigarette.*

Sid Richardson Collection of Western Art, Fort Worth, Texas

The actual details of Jake's discovery have gone unrecorded; the tales that survive are obviously romanticized. The most common version finds Jake not prospecting at all, but seeking shelter from a sudden storm. Crawling beneath a ledge, a piece of rock dislodges in Jake's hand. Inspecting it, Jake finds nothing less than a museum-grade specimen of lode gold. In the popular mining lore of the West, most gold discoveries are somehow tied to luck and coincidence. If a fraction of those tales were true, more gold would have been discovered by bullets ricocheting off ledges and deer kicking up dirt than was ever found by the drudgery of manual labor. In a much more likely scenario, Jake Hoover packed into lower Yogo Gulch with his grub and gear, intent upon methodically prospecting the gravels. Cursing and sweating, he dug many times to bedrock, panning samples as he went. In one of those holes, Jake panned enough color to interest him. As he kneeled in lower Yogo Gulch examining the black sands and tiny bits of color, he also noticed a number of tiny, translucent blue pebbles. Just as other prospectors had done fifteen years before, Jake rolled them around in his fingers, studied their unusual color and clarity, then casually dropped them back in the creek and resumed his search for gold.

Hoover realized that proving the richness and extent of his modest gold strike was beyond picks, pans and a trickle of water. Seeking assistance, Hoover visited his old friend, S. S. Hobson, a successful rancher and president of the Fergus County Bank in Lewistown. Hobson was interested in the strike and contacted a friend, Dr. Jim Bouvet, a Chicago veterinarian. During the winter, Bouvet traveled to Montana and was also favorably impressed with Hoover's find. By spring, 1895, an equal partnership was established between Hoover, Hobson and Bouvet, with Bouvet providing the bulk of the $40,000 operating capital.

A Butte engineer was hired to construct a ten-mile-long ditch and flume system to divert water from upper Yogo Creek to the lower workings. The job took only two months, but devoured $38,000 of the partnership funds. The remaining capital was just enough to build the sluices and hire a few local laborers whom Hoover supervised through the summer. But at season's end, Jake found his golden dream had eluded him once again. With the final cleanup of the sluices, the partnership's accumulated gold barely filled a coffee cup. It amounted to some forty Troy ounces worth a bit over $700. The gold mining venture in lower Yogo Creek had been a financial disaster.

With little gold to worry about, Jake had busied himself during that long summer with collecting the blue pebbles that appeared in the sluice concentrates—the first recorded mention of Yogo sapphires. Jake Hoover's place in history was now assured; not only was he Charlie Russell's friend, he was now the discoverer of Yogo sapphires. It is uncertain if Hoover really knew what the blue pebbles were, or what his motives were for collecting them. Just as his gold discovery became steeped in romance, so, too, did his sapphire discovery. One published version finds Jake walking down Yogo Creek and tripping over "a sapphire bigger than a hen's egg." In another, Jake kneels to drink at a clear pool, and his eye is "drawn to a mote of blue, bluer even than the sky above and glowing like a hot coal." Plucking this "mesmerizing object" from the water, Jake finds "a sapphire jewel as big as his thumb." A slightly more plausible, but equally erroneous, account has Jake believing the blue pebbles are broken whiskey bottles that had been abraded and rounded in the creek gravels. Fifteen years of sporadic mining in the Yogo district had doubtlessly put many broken whiskey bottles in Yogo Creek, and Jake may have deposited a few there himself.

At the end of the 1895 mining season, Jake Hoover had collected enough of the blue pebbles to fill a cigar box. Most tales have Jake wondering naively what the blue pebbles were. Depending upon the particular version, Jake was finally told by either Hobson, a Great Falls lapidary, or a Maine schoolteacher that the blue pebbles were sapphires. Jake's comment then, supposedly, was, "What in hell is a sapphire?"

Jake Hoover had spent too many years in the gold camps around Helena not to know what a sapphire was. Nor did he think the blue pebbles were bits of broken whiskey bottles. Jake collected the blue pebbles because he suspected they might have value. After being told the blue pebbles were sapphires, Jake Hoover never asked, "What in hell is a sapphire?" He probably asked, "What in hell are they worth?"

Hoover, Hobson and Bouvet soon learned what they were worth. The box of blue pebbles, along with samples of the sluice concentrates, was mailed off for identification and appraisal. After going through at least one assay office that offered only an inconclusive evaluation, the box reached New York City and the offices of Tiffany & Co., where it was soon placed upon the desk of none other than the foremost gem expert in the United States, Dr. George Frederick Kunz.

Kunz was indeed interested in the package that had arrived from Lewistown, Montana. A few simple tests of specific gravity, refraction and hardness quickly confirmed Kunz's suspicion that the blue pebbles were sapphire. But it was the quality and color that fascinated the gemologist. Unlike the countless other Montana sapphires Kunz had examined, which were so difficult to market because of their colors, nearly every one of these fell within a tight range of very desirable blues. Kunz knew immediately that, in quality and color, these were the finest precious gemstones ever found in the United States.

A few of these first "Yogos" were cut to reveal their full beauty. There were almost none of the usual inclusions, the flaws, that detracted from beauty and "cuttability." The color was remarkably even, not zonated, as in oriental sapphires. And the Yogos exhibited life and brilliance not only in sunlight, but even in artificial light,

George Frederick Kunz was among the first to recognize the beauty of the Yogo sapphire and the commercial potential of the Yogo dike. Many of the Yogos collected by Kunz became a part of the J. Pierpont Morgan Collection which is now in the American Museum of Natural History in New York City. In 1895, Kunz was the gemologist with Tiffany & Co., who identified the cigar box full of blue pebbles collected by Jake Hoover.

when most foreign sapphires turned dark and lifeless.

That cigar box of blue pebbles never returned to Montana. In its place came a check for $3,750, along with a letter from Tiffany & Co. declaring the stones to be "sapphires of unusual quality." The amount of the check had been based upon London market prices for rough sapphires of $6 per carat for stones of first quality, $1.25 for stones of second quality, and 25¢ per carat for the gleanings.

In his reminiscences, which appeared as installments in 1927 and 1928 in *The Saturday Evening Post*, George Frederick Kunz recalled receiving that cigar box full of Yogo sapphires.

> To the layman, it would perhaps seem improbable that, sitting at a desk in New York, one should be able to discover a gem mine in Montana; yet that is just what happened to me once.
>
> One day some years ago I received by mail—which reminds me that the Cullinan diamond, which one would naturally expect would journey under armed guard, was sent for safety all the way from the Premier Mines in South Africa to London by ordinary registered mail—one day I received some specimens in which occurred grains of gold. These had been mined in Yogo Gulch, Montana, on a property bought as a gold mine. As a matter of fact, very little gold was ever located there. But upon examination, I found certain crystals to which little attention had been paid, but which, on examination, I discovered to be fine blue sapphires. When this information was conveyed to the owners they immediately began to search for other specimens, and soon brought to light the fact that, in purchasing a mere gold mine, they had acquired the most valuable sapphire mine in America, yielding more wealth than all the other sapphire mines in America put together—and a finer quality of gem.

When the Tiffany & Co. letter reached Montana, the thoughts of Hoover, Hobson and Bouvet quickly turned from gold to sapphires. There was some question whether sluicing sapphires from lower Yogo Creek would ever recoup their original investment. They never found out because of another discovery, this one made by a local sheepherder.

In February, 1896, Jim Ettien was herding sheep on the gently rolling bench lands just east of Yogo Gulch and Tollgate Hill when his attention was drawn to gopher heaps. As a Montana sheepherder, Ettien had seen his share of gopher diggings but, somehow, these were different; instead of being randomly scattered, they were arranged in a line that cut arrow-straight across the hills. Investigating, Ettien found that the earth within that narrow line was much softer than that which thinly covered the surrounding limestone bedrock. He followed the slight, grassy depression, conveniently marked by gopher diggings, nearly a mile; in some places it was twenty feet wide, while almost pinching out to nothing in others. Ettien, having done a bit of prospecting himself, surmised it to be some sort of vein, possibly a mineralized vein. As any other Montana prospector would have done, Jim Ettien guessed that this strange geological formation just might bear gold. The next day, Ettien returned with a wagon, shoveled some of the soft earth from the gopher heaps into sacks, and hauled it down to Yogo Creek for washing. Instead of gold, Ettien found blue pebbles—the same blue pebbles that Jake Hoover and his partners had washed out of lower Yogo Creek the past mining season. Ettien knew they were sapphires and, from Hoover's experience, quite valuable. On his next trip to Lewistown, Ettien filed two lode claims on his "vein" that would soon become known as the Yogo dike. For the next few weeks the sheepherder Jim Ettien held sole ownership to what would soon be recognized as North America's largest and richest deposit of precious gemstones.

Word of Jim Ettien's sapphire claims on the bench lands just east of Yogo Creek soon reached Hoover and his partners. By April, enough snow had melted for a good look at the area. Hoover found the faint depression marked with the gopher diggings knifing across Ettien's two claims—and continuing west about two miles, right to Yogo Creek itself. Jake was hardly a geologist, but enough of a practical prospector to put things in perspective. The sapphires he had washed out of Yogo Creek had come from a secondary deposit. The creek had cut through the dike, washing some sapphires downstream and concentrating them with the gold. The dike itself was the lode source of the sapphires. The next day, Jake Hoover was in Lewistown filing eight lode claims in the name of his partnership.

The first mining of the Yogo dike took place in summer, 1896. On the surface, the dike rock had become weathered into a soft, crumbly clay ready for washing; at depths beyond fifteen feet, however, the dike rock was found to be solid and unyielding. Mining was limited to shallow digging. Since there was no available water on the bench lands, the "dirt" was loaded onto wagons and hauled down to the Judith River for washing. By September, Hoover, Hobson and Bouvet had recovered several thousand carats of rough sapphires.

A mining engineer observed that the location of the two Ettien claims could create future problems. Situated between the partnership's property and the Judith River, they interfered with access and, if water were ever brought to the bench lands for mine site washing, would also interfere with tailings disposal. Jim Ettien did not mine his claims in 1896 and apparently had no plans to begin. He was troubled both by the high cost of bringing water to his claims and the question of where to sell his sapphires after mining them. It was also said the fabled oriental sapphires were all mined from placer deposits, and that the Yogo dike was the only lode sapphire deposit known in the world. Ettien did not even know if lode sapphire mining was economically realistic; he did know that no one in the state of Montana had yet to make a nickel mining the gemstones. The Hoover-Hobson-Bouvet partnership, meanwhile, had been advised that future mining development along the dike hinged on acquisition of the Ettien claims. When notified of the partnership's offer to purchase his property, Ettien didn't give it much thought. In fall, 1896, Jim Ettien sold his two claims on the Yogo dike for $2,450.

Jake Hoover, S. S. Hobson and Jim Bouvet now controlled about two and one-half miles of the Yogo dike, most of its known length. Although their early mining was exploratory and experimental in nature, Yogo sapphires began receiving some of their first publicity. In an 1897 issue of the *American Journal of Science*, Dr. George Frederick Kunz compared the Yogo sapphires with those found earlier in Montana:

> As to the value of the early Montana sapphires in jewelry, it is hardly possible to predict how far it may really be important. Much beautiful material has already been obtained, but little of high value. Those from the Missouri bars had a wide range of color,—light blue, blue-green, green and pink, of great delicacy and brilliancy, but not the deep shades of blue and red that are in demand for fine jewelry. As semi-precious, or "fancy" stones, they have value, however.
>
> The Yogo Gulch-Judith River region is more promising, the colors varying from light blue to quite dark blue, including some of the true "cornflower" blue tint so much prized in the sapphires of Ceylon. Others incline to amethystine and almost ruby shades. Some of them are "peacock blue" and some dichroic, showing a deeper tint in one direction than in another; and some of the "cornflower" gems are equal to any of the Ceylonese, which they strongly resemble,—more than they do those of the Cashmere. Several thousand carats were taken out in 1896, from a preliminary washing of 100 loads of the "earth": of these, two hundred carats were of gem quality and yielded, when cut, sixty carats of fine stones worth from $2 to $15 per carat. All, however, are small, none having yet been obtained of more than 1½ carats in weight.

In 1897, the death of Jim Bouvet and the need for additional operating capital prompted a reorganization of the original partnership into the New Mine Sapphire Syndicate, incorporated under the laws of the State of Montana, with an authorized capitalization of $100,000. The two new partners were both residents of Great Falls; Mathew Dunn was a businessman and politician, George A. Wells was an Englishman with limited background in gems. Matt Dunn was installed as president, Hobson as vice president and George Wells as secretary. Along with Jake Hoover, all were equal partners. The name of the company, the New Mine Sapphire Syndicate, may have been originated by Wells to imitate the names of the leading British gem "syndicates."

From the beginning, it was obvious that rough Montana sapphires could never be sold at a profit. The roughs would have to be cut, then marketed as finished gems. Both the cutting and marketing required much more expertise and experience than was available in Montana. Accordingly, George Wells traveled to New York to raise additional capital and to test market interest for the Yogo sapphire. His first stop was Tiffany & Co., where the Yogos had received their first favorable appraisal. But neither Tiffany nor any other New York gem merchant Wells contacted had much interest in becoming marketing agents for the future mine production. The American reluctance in supporting the Yogo sapphires was understandable; attempting to buck the established, British-controlled gem industry with yet unaccepted sapphires from a yet unproven source entailed considerable risk. Besides, the image of Montana sapphires was one tarnished by fraud and stones of unmarketable colors.

Disappointed but not discouraged, Wells then sailed to London to continue his efforts. Among his first stops were the offices of Johnson, Walker and Tolhurst, Ltd., one of England's most distinguished gem merchants, jewelry manufacturers and wholesale jewelers. After displaying his samples of rough and cut Yogo sapphire, Wells knew he had come to the right place. Since watching De Beers take control of the South African diamond fields, and thus the bulk of the world diamond market, Johnson, Walker and Tolhurst had contemplated a similar move in sapphires and had already acquired substantial gem mining interests in Ceylon and Burma. George Wells and his Yogo sapphires represented opportunity walking right in the front door. Johnson, Walker and Tolhurst quickly agreed to a ten-year contract as the worldwide sole marketing agents for Yogo sapphires. So great was the British interest, the directors even persuaded Wells to sell some of his shares in the New Mine Sapphire Syndicate.

While the British were acquiring a minor interest in the New Mine Sapphire Syndicate in London, further change was coming to the Syndicate in Montana. Jake Hoover, already forty-eight years old, was thinking of moving on for a variety of reasons. Jake had married in 1894, but his problems with women continued. The marriage had gone from bad to worse, progressing, as Jake's matters with women always did, to settlement with a Winchester rifle. This time, his wife's finger was on the trigger and Jake was staring down the bore. The hot lead, however, was a bit off its mark, whistling by Jake's ear. By his own admission, it was the closest he ever came to death.

Since coming to the Judith River Basin nearly two decades earlier, Jake had always remained in debt to the T. C. Power & Brother trading company at Fort Benton. There was no reason to ever believe that debt would be erased; Jake's income was limited to his hunting, his assets to his small ranch and a quarter interest in a sapphire mine. But, in the three years since he had dug those blue pebbles out of Yogo Creek, sapphires hadn't brought him a cent. It wasn't sapphires, but gold that weighed heavily upon Jake Hoover's mind in 1897. Like every other Montana prospector and miner, he had heard the tales of the great gold strike on the Klondike River in Canada's Yukon Territory. The rush to the North had already begun in the ports of San Francisco and Seattle. Jake knew that life in the gold fields was best suited to younger men, so if he was ever going to find his golden bonanza, he'd better do it soon. Jake could even see that Montana was changing; the once wide open prairies were already crisscrossed with telegraph poles, iron rails and barbed wire. Even the game wasn't as plentiful any more, and life itself was becoming more complex. The company in which he owned a quarter-interest was, at that very moment, engaged in mysterious, involved negotiations with "some English outfit" about cutting and marketing, matters that Jake neither understood nor really cared about. He knew, too, that his roughhewn way didn't fit with those of the educated, successful businessmen who were now his partners. Jake Hoover wanted, and needed, a return to the simple world of the gold pan and rifle.

Jake Hoover went to his old friend S. S. Hobson and told him he wanted to sell out. On behalf of the Syndicate, Hobson offered Jake

The only known photograph of Jake Hoover, taken in 1894 at Lewistown when Hoover was forty-five years of age. The "soft suit of clothes" had been ordered to wear at his wedding. Hoover was remembered as a friend of Charlie Russell, as a hunter whose shooting skills were legendary, and as the man who discovered the "blue pebbles"–Yogo sapphires–in the gravels of Yogo Creek.

$5,000 for his quarter-interest. Jake took it, paid off his debt to T. C. Power & Brother, put his ranch up for sale, and began straightening out his personal affairs in preparation for eventual departure to the new gold fields in the North. No one yet knew what that dike full of sapphires was worth, but the two men who had the most to do with finding it, Jim Ettien and Jake Hoover, wouldn't have to worry about it.

In July, George Wells returned to Montana with two representatives of Johnson, Walker and Tolhurst, Ltd., who inspected the Yogo dike. What followed was the first real indication of the value of the

New Mine Sapphire Syndicate shares. The *Great Falls Tribune* carried the story on August 24, 1897.

THE OPTION TAKEN UP
A QUARTER INTEREST IN
THE YOGO MINE AT YOGO
SOLD TO ENGLISHMEN

THE DEAL WILL RESULT
IN GREATER ACTIVITY
IN OPERATIONS
IN THE NEXT YEAR

Messrs George A. Wells and Mathew Dunn of this city yesterday sold to Mssrs Johnson, Walker and Tolhurst, capitalists of London, a one-fourth interest in their sapphire mines at Yogo, the price being over $100,000.

A few weeks ago Mr. Tolhurst and Edward A. Keller, a stone broker of London, visited the mines in company with Mr. Wells, who had shown them some specimen stones while in London, and were well pleased with the property.

They wired the result of their examination into London and the gentlemen who have become purchasers secured an option on a quarter interest in the property, which would have expired on September 15. They were so well satisfied with the full reports on the property that they decided at once to take up the option and did so.

It is probable that the other three-fourths interest in the sapphire fields will be placed upon the market in the near future.

Mssrs Wells and Dunn have expended $60,000 in developing the sapphire fields and have established a world-wide market for the stones, which rank among the best ever produced. The field is the only one in the state which has proved its value.

The sale will next year result in greater

> activity in the mines and their development will be watched with interest by the people of northern Montana. Mr. Wells stated that the present season will end October 1 and that the work will then be suspended, to be resumed on a larger scale next spring.

Jake Hoover had sold his quarter-interest in the New Mine Sapphire Syndicate for $5,000, and was probably satisfied that it was a fair price. Only three months later, Johnson, Walker and Tolhurst, Ltd., paid over $100,000 for exactly the same interest. The timing had been ironic; once again, Jake Hoover had demonstrated his unique ability to discover mineral wealth, yet never profit from it. Many years later, Jake would confess, in a classic understatement, "I have always had poor luck in that I always sold out at the wrong time."

Before the involvement of Johnson, Walker and Tolhurst, the Yogo dike, under American management, had produced only a few thousand carats per year, only a small fraction of which were of gem quality, and had yet to make a financial return. In 1899, British mining expertise arrived in the person of T. Hamilton Walker, the son of one of the principals of Johnson, Walker and Tolhurst. Walker, a mining engineer with De Beers, had been recalled from the South African diamond fields and sent to Yogo to increase production. Improvements were made to the twelve-mile-long ditch carrying water to the mine from upper Yogo Creek and a small steam plant was constructed to power the pumps for hydraulic mining. Hydraulicking was the perfect system for mining the uppermost, exposed and weathered sections of the dike rock. High pressure water jets were used to erode the weathered dike rock away, turning it into a mud that was passed through a series of sluice boxes set directly in the dike at a slightly lower gradient. By October, when the approaching winter froze the water supply, the dike had yielded 400,000 carats of sapphires, 125,000 of which were of gem quality. The gem material, of course, was shipped to Johnson, Walker and Tolhurst, Ltd., for sorting and cutting. The non-gem sapphires, the stones that were too small, too flat, off-color, or otherwise imperfect, were sold to industry for use as mechanical jewel bearings or abrasives. The value of the

An early photograph looking northward up Yogo Gulch. Taken about 1900, the photograph shows the road and the water flume bringing water from upper Yogo Creek to the English Mine. Yogo city is about three miles up the gulch.

1899 mine production at the Yogo dike was about $70,000, with three-fourths of that sum accounted for by the gem material.

While the Yogos were decidedly smaller and flatter than most oriental sapphires, their quality and fine color assured them of a place in the European markets. The future seemed encouraging, yet the Americans involved in the Yogo dike showed little interest in participating in the development of the mine and the profits to come. In 1899, Mathew Dunn sold his share to the British. Then, in 1901, S. S. Hobson, now a Montana state representative and the last remaining American shareholder, did the same. The British now had complete control over the New Mine Sapphire Syndicate, most of the Yogo dike and the marketing of Yogo sapphires.

The steadily developing mine was twelve miles way from Utica, the nearest town and stagecoach stop. Many of the crew of twenty workers chose to live in the growing collection of small cabins and bunk-

The earliest known photograph of mining at the Yogo dike, taken about 1897. In this hydraulicking operation, the high-pressure hose is used to wash way the weathered surface dike rock into sluice boxes. The trench in which the hose lies is the dike itself. Water was brought by a ditch and flume system from upper Yogo Creek to the top of the hill in the background. Water pressure was gained by the 150-foot "head," or drop in elevation from the hill to the dike.

houses at the mine site. The United States Post Office already unofficially recognized the neophyte community as "Sapphire, Montana."

Five miles away in tne ruins of Yogo City on upper Yogo Creek, Millie Ringgold still clung stubbornly to her fading dreams. Old Millie's dreams would never be more than just that, for the only mining news ever again to come out of the Yogo mining district would come not from Yogo City, but from that long, narrow, sapphire-bearing dike that was already becoming known as the English Mine.

Chapter 2

Sapphire!

Sapphires, although relatively new to the Montana miners of the late nineteenth century, were among the first gemstones known to man. During the 10,000 years that man has collected sapphires, he has associated those precious blue gemstones with an enormous amount of rich lore and legend which, only in recent times, has been tempered with scientific understanding.

The first gemstones man collected may have been bits of water-worn, semi-polished quartz from the sands of a primal beach. Although worthless in the utilitarian sense, such pebbles had a beautiful luster, brilliance and color to distinguish them from common gravels. As early man searched for stones for tools and weapons, he discovered other forms of mineral beauty—natural mineral crystals which, with their orderly structure, straight lines and smooth planes, must have seemed a miracle of creation. Even more remarkable was a clarity that permitted one to see within the stone to a mystical world of reflection and refraction that magically trapped the sunlight and brought it to life. Besides beauty, many gemstones had another desirable quality, that of permanence. Unlike that of flowers and sunsets, the beauty of certain gemstones neither faded nor deteriorated; it was eternal, outlasting man, and generations of men. Many minerals were beautiful, but rarely were they both beautiful and durable. It was, therefore, very early in man's experience with gemstones that he had defined their three basic requisites: *beauty*, *durability* and *rarity*.

Gems were destined to have a very broad appeal and serve many purposes for man. Like gold, they would become the gifts and ransoms of royalty, objects of sacrifice, and the causes of theft, violence and murder. Gems would be used as symbols of power and authority, as stores of wealth, and as gifts given out of the conflicting motivations of fear and love. The wearing of gems began as the simple desire to display a distinguishing mark that would command respect and attention. Certain prized gems became the royal insignia of secular power; soon men were attributing supernatural powers to gems that, hopefully, might give them control over natural forces. Gems became selectively worn as amulets, to bestow good powers and qualities, and as talismans, to negate and absorb evil powers. From there, it was inevitable that gems would play important roles in religion and medicine.

Some civilizations would become associated with the gemstones they most revered and best worked: turquoise for the Aztec and Egyptian, jade for the Chinese, and rubies for the Hindus. By the time of Christ, most of the minerals we now recognize as gemstones had already been discovered. By the mid-1700s, after worldwide exploration and commerce had greatly expanded our knowledge of gems and their sources, only four gemstones retained a universal recognition and acceptance as truly precious. These stones represented the ultimate in beauty, durability and rarity. They were the diamond, the emerald, the ruby and the sapphire.

Ruby and sapphire, both gem forms of the mineral corundum, were the first precious gemstones to be discovered and mined. Stone age relics dating to 8000 B.C. have been recovered from the ruby and sapphire mines of Burma. Man soon learned that these two gemstones, with their rich ruby reds and deep sapphire blues, were the hardest materials yet found, and therefore among the most durable. A thousand years before Christ, sapphires and rubies were prized commodities in the developing spice trade between India and the Mediterranean. Ancient writings of the spice trade mention corundum, a word which may be traced back to the Tamil term *kurundam*, or ruby. Many writings in Sanskrit, an early language of India, tell of a great reverence and demand for rubies and sapphires. Our modern names for these gemstones derive from the Latin *rubeus*, or red, and from the Greek *sapphirus*, meaning blue. Originally, the word sapphire referred to lapis lazuli, an opaque, blue decorative stone. Only since the Middle Ages has the name been assigned to the blue corundum gemstone. Much ancient lore first connected to lapis lazuli and the color blue has since become associated with true sapphire.

To many early astrologers, the subtle changes of hue, tone and brilliance seen when a gem was turned in the fingers were signs that foretold the future. These mystics found color the most important quality of a gem, and classified stones as positive or negative accordingly. Warm colors such as reds and yellows were thought to bestow good powers, while the negative colors, blues and greens, were believed to absorb poison and misfortune. Thus, ruby and sapphire, which are the same basic mineral, grew far apart in concept and lore.

Although the rudiments of gemstone mythology had appeared in prehistory, it was greatly advanced by the jeweled breastplate of Aaron, the first high priest of the Temple of Jerusalem. As described in *Exodus* 28:17-20, there were twelve gems, arranged in four horizontal rows of three stones each, that represented the twelve tribes of Israel. The first breastplate was made thirteen centuries before Christ, but was lost during the captivity of the Jews by the Babylonians. After the Kingdom of the Jews was restored, about 500 B.C., a second breastplate was made, which most scholars feel did not exactly duplicate the original. Numerous language changes had confused the identity of the original stones. Meanwhile, new gemstone discoveries, along with advanced cutting and polishing methods, offered new gemstone choices. The actual stones used will probably never be known, for the second breastplate disappeared about five centuries after Christ.

The gems used in the breastplate—more specifically their particular position within the breastplate—became associated with many fanciful religious ideas. The supposed powers of the breastplate were many; changes in gem color or brilliance could signal everything from the approach of death to victory in battle, even to the presence of God himself. According to ancient descriptions, ruby was the right hand stone in the second row from the top, sapphire was to its immediate left. It is doubtful that the corundum gems—the sapphire and ruby we know today—were actually used in the breastplate. Probably, the breastplate's "sapphire" was lapis lazuli. Even though the breastplates were lost, many of the beliefs tied to the names of the stones survived; by the Middle Ages, they had evolved into a complex mass of superstitions and dark beliefs of hoped-for planetary, medical and alchemical powers.

Among the earliest books discussing gems were the *Peri Lithon* (Of Stones), written about 300 B.C., and, more importantly, Pliny's *Historia Naturalis*, written about A.D. 50. Although neither could be considered scientific, names were given to minerals and distinctions made between them. For over 1,000 years, *Historia Naturalis* was the sole basis for mineral classification, yet it was a model of metaphysics, a mix of parochial reality and myth that would reach its extreme in the Middle Ages. With the limited scientific thought of this period, the myths surrounding gems, instead of being dispelled, grew enormously. Every tale, regardless of its source of plausibility, was written as fact and often became the basis for alchemical experimentation. Gemstones were assigned a variety of magical powers, more or less in keeping with the ancient lore.

Sapphires, with their purported power to absorb poisons and evil, were widely employed by medieval physicians. Elaborate potions were concocted, duly recorded in the pharmacopoeias of the age, and handed down to the next generation for improvement. Two of the most popular sapphire-based remedies were the *Electaurium ex Gemmis Johannis Mensual* and the *Confectio Hyacinthi*. Both were complex, yet basically similar; a modern pharmacist might consider them a broad-spectrum drug, for they contained not only the medical benefits of ground sapphire, but pearls, garnets, emerald, flakes of gold and silver, and various herbs. But so great were considered the powers of the sapphire, that the gem was often administered alone. The stones were crushed to a powder and prepared as a paste which, thankfully, facilitated both swallowing and passage throgh the digestive tract. A common dose was "ten to forty grains" of ground sapphire. Today, the maximum dose would be rather expensive, for it contained the rough equivalent of a twelve-carat stone.

At best, powdered sapphire, or aluminum oxide, may have had a mild antacid effect. Although it had no true curative powers whatever, it was a favorite of the most esteemed physicians. The following excerpts from the writings of early physicians show the sapphire's place in well over three centuries of medieval medicine.

> Sapphire is medicine; for, being powdered it heals the sores following pustules, and boils, if smeared over them, being thus applied mixed with milk to the ulcerations.

The vertues of the Saphyre are—it is cold, dry and stringent: it dryes up rhuems in the eyes and takes away their inflammation, being used in collysiums, or to anoint the eyelids. It is good in all fluxes of the belly, the dysentary, the hepatick flux, the haemorrhoids, and other bleedings; it cures internal ulcers, and wounds, strengthens the heart, and refreshes it; it is an enemy to all poisons; it likewise causes melancholy.

A Sapphyre, or a stone that is of a deep blue color, it be rubbed on a tumour wherein the plague discovers itself (before the party is too far gone); and if, bye and bye, it can be removed from the sick, the absent jewel extracts all the poison, or contagion therefrom.

Its virtue is contrary to venom, and quencheth it at every deal. And if thou put an addercop (viper) in a box, and hold a very Sapphyre of Ind (India) at the mouth of the box, and while, by virtue thereof the addercop is overcome, and dieth as if it were suddenly. And this same I have seen oft proved in many and diverse places.

The Sapphire heals sores; it is found to discharge a carbuncle with a single touch.

Since the blue of many sapphires resembled that of the iris in some human eyes, physicians believed the sapphire was of particular ophthalmic benefit.

A whole Stone laid to the forehead stops bleeding at the nose, and when applied to inflammations, abates them. Being brought into little balls, as big as peas, and polished, and put in the eyes, it takes out anything that is fallen in, dust and gnats; and preserves the eyes from the small pox, and other diseases. A Saphyre is prepared the common way, by levigation, with cordial water. Others dissolve the fine dust of a Saphyre in pure vinegar, and the juyce of limons, and give the solution, with some other cordial.

A classic tale of the ophthalmic benefit of sapphires was recorded in *Gesta Romanorum* in 1362. The blind Roman emperor Theodius ordained that the plight of any aggrieved person could be brought to his immediate attention by ringing a bell in his palace. The bell rope hung near a serpent's nest. One day, in the serpent's absence, a toad occupied the nest. Upon returning, the serpent coiled itself about the rope and rang the bell, demanding its own justice. Hearing of this, Theodius ordered the toad killed. Later, when the emperor slept, the serpent entered his chamber. In its mouth was a large sapphire which it placed upon the emperor's blind eyes. From that moment on, the sight of Theodius was said to have been restored.

In London, in 1391, Richard Preston offered a sapphire as an "eyestone" at the shrine of St. Erkinwald at Old St. Paul's Cathedral. Preston stipulated that the stone must always be kept at the shrine and made available to anyone for the curing of eye ailments. Sapphires soon began appearing in other European cathedrals for the same purpose.

Sapphires were not limited to physical medicine, but also worked upon maladies of the mind and soul. While ruby was related to earthly matters, sapphire, possibly because of its sky- or heaven-like color, was related to qualities of a "higher" nature. Spiritualists believed sapphires aided in hearing voices from the beyond and in understanding "the most obscure oracles" from the spirits. Sapphire became a symbol of chastity, love and trust, a stone that encouraged faithfulness, lifelong devotion and tried affection. In later times, this symbolism would make the sapphire appropriate for rings of betrothal.

Ever since the ancient Persians explained the blue of the sky by believing the earth rested upon a great sapphire, the stone has been associated with the heavens, a natural basis for its later ecclesiastical significance. A sixth century Papal Bull ordained that all Cardinals, as a sign of faithfulness to the Church, would wear a sapphire ring on their blessing hand. Sapphire came into even greater favor with the Church in the twelfth century when the Bishop of Rennes lauded the stone for its beauty and power. Soon afterward, sapphire became a pontifical stone, and the Church's most significant item of jewelry was a gold ring set with a blue sapphire.

Even as late as the great age of exploration and discovery, sapphires and many other gemstones were thought to be composed of "celestial matter." Their powers and influences were assumed without question. But the spirit of investigation of the Renaissance had already brought many aspects of the natural world, including gemstones, under serious study, The early scientists did not dispute the accepted gem beliefs; instead, they tried to learn why gems could modify the health, character and fortunes of the wearer. These attempts to explain the nonexistent power of gems were the first feeble steps of the neophyte science of gemology.

In *De Mineralibus et Rebus Metallicum*, Albertus Magnus agreed that gems doubtlessly possessed "celestial virtues," but might not really have spirits of their own. Georg Bauer, writing under the pen name Agricola, established a mineral classification system that tied gems not to celestial matter, but to other common rocks and minerals. Later, in 1663, Hieronymus Cardenus speculated that the shapes of gemstones could be attributed to certain natural forces that controlled crystalline growth. In the same year, Sir Robert Boyle, the English physicist, noted that his studies had done little to prove the power of gems.

> But what I chiefly consider is on this occasion that 'tis one thing to make the possibility probable that gold, rubys, and sapphyrs, etc., may be wrought upon by the human stomack; and another thing to show both that they are wont to be so, and that they are actually endowed with those particular and specifick vertues that are ascribed to them.

A few years later, Boyle advanced the astounding proposition that ruby and sapphire, for all their wide differences in color, lore, supposed powers and symbolisms, might actually be the same mineral. Boyle noted, "the degree of hardness of rubies and sapphires is so equal that a jeweler takes them to be the same stone except for the colour." Boyle even demonstrated that their densities were equal.

Although science would soon provide answers for the mysteries of gemstones, belief in the celestial powers and astrological influences of gems would survive. Both ruby and sapphire had obvious planetary symbolism, ruby appearing much like the red planet Mars and the cool blues of sapphire telling of Venus. The basis of our modern concept of monthly birthstones, which originated in Poland in the nineteenth century, may be traced back to the biblical breastplates of twelve stones, each with its individual significances. George Frederick Kunz, in his *The Curious Lore of Precious Stones* (1913), mentions sapphire as "the gem of autumn, the blue of the autumn sky . . . is therefore appropriate to the autumn season, when the declining sun no longer sends forth the fiery rays of summer but shines with a tempered brilliancy." Sapphire, accordingly, was considered appropriate to, and symbolized the influences of, the month of September.

By 1800, great European advances in chemistry, physics and the allied science of mineralogy had begun stripping the shroud of superstition and myth that surrounded gemstones, replacing it with hard scientific knowledge and understanding. Rather than "celestial matter," precious gemstones were found to be made of common elements. Diamond, for example, was composed of carbon; rubies and sapphires of aluminum and oxygen. Unlike gold, which had an intrinsic value, gemstones owed their value to beauty and rarity of form. Although the earth had over 2,000 minerals, only a few dozen possessed the attributes to qualify as gemstones. And classic, truly precious gemstones numbered only four. Ancient lore had bestowed upon gemstones an enduring mystique. Now, the emerging science of gemology would prove no less intriguing.

Both sapphires and rubies were found to be a transparent, crystalline form of the mineral corundum, or aluminum oxide (Al_2O_3). The internal arrangement of the two atoms of aluminum and three atoms of oxygen determined the basic external hexagonal, or six-sided, shape of the corundum crystal. All gemstone qualities, from durability to how they affect the passage of light, were found to depend upon the particular chemical composition and internal atomic structure.

After color and brilliance, the next quality of the sapphire to be recognized was extreme hardness, or the ability to resist scratching. Before 600 B.C., when diamonds were discovered, the corundum gems were the hardest material known to man. Sapphires and rubies had an early utilitarian use, that of engraving and crudely shaping other gemstones. It was not until the early 1800s that the German mineralogist Friedrich Mohs created a comparative scale of mineral hardness ranging numerically from 1 to 10. Talc (1), the softest mineral, was followed by gypsum (2), calcite (3), fluorite (4), apatite (5), feldspar (6), quartz (7) and topaz (8). Corundum, with only slight variations between sapphire and ruby, ranked at 9. Diamond, the hardest natural material, completed the scale at 10. The difference in hardness between numbers was not at all uniform; greater difference existed between sapphire (9) and diamond (10) than between the entire remainder of the Mohs scale. Any mineral that could scratch any other mineral received a higher numerical ranking. Common materials that lend the Mohs scale a practical perspective are fingernail (2½), copper penny (3), window glass (6) and alloyed steel files (7).

The most common abrasive we encounter every day are fine particles of quartz (7), the major component of dust. Generally, mineral crystals softer than 7 will not withstand the wear encountered in regular jewelry use. Rhodochrosite, for example, a rose-red, crystalline form of the mineral manganese carbonate which rivals the red of the best ruby, has a hardness of only 4—unsuitable for regular use. Sapphire and ruby, harder than any other mineral except diamond, are particularly well suited for jewelry use.

Toughness, a quality of durability often confused with hardness, is the ability to resist impact that might cause chipping or even shattering. An analogy would be glass (6) and copper (3); an impact that would shatter glass would not damage copper. In toughness, sapphires rank extremely high.

The first measurement of precious stones was one of simple weight. In many ancient cultures, small units of weight were measured in their equivalent of carob seeds, a widely cultivated seed that was remarkably uniform in weight. The carob seed is the origin of our "grain" unit of weight. The origin of today's carat, the international unit of gem weight, also stems from the carob seed, which was *qirat* in Old Arabic and *carato* in early Italian. For centuries, the carat system was quite confusing, for the commmonly used French, British and Indian carats were all of significantly different weight. There was additional confusion between the carat and the *karat,* which had nothing whatever to do with gems, but was a twenty-four-part system used to define the purity of gold. In 1913, all gem weights were standardized with creation of the international carat, weighing one-fifth of a gram, and containing 100 points.

Another important gem measurement is density, or weight per unit volume, which is expressed as specific gravity. The specific gravity of sapphire is about 4.0, meaning it is four times heavier than an equal volume of water. In comparison, quartz, and most other common rocks and minerals, have specific gravities of only about 2.6. This considerable weight differential explains the natural alluvial concentration of sapphires, and also how the Montana miners were able to recover the gemstones in sluice boxes, just as they recovered bits of heavy placer gold.

Since different gemstones vary considerably in density, the sizes of different stones of equal weight will also vary. A one-carat diamond, for example, will be about fifteen percent larger than a one-carat sapphire, for diamond's specific gravity is only about 3.5.

The grandest property of a gem is neither weight nor hardness, but the manner in which it affects light to produce color, luster and brilliance. Color, or the absence of color, as well as the degree of transparency, are functions of light absorbency. If all the wave lengths of white light striking a gem are transmitted or reflected, the stone will appear clear and colorless. If all the wave lengths of white light are absorbed, the stone will appear black. Partial absorption of white light, however, will produce colors. If the blues and greens are absorbed, a gem will appear yellow or red; conversely, absorption of the red and yellow wave lengths will "color" a gem blue or green.

All colored gemstones may be classifed as either idiochromatic or allochromatic. The colors of idiochromatic gemstones depend only upon chemical composition and crystalline structure. Peridot is idiochromatic, for its color must always be green, just as rhodochrosite must always be rose-red. Allochromatic gemstones, however, owe their colors to the presence of trace elements acting as coloring agents, or chromophores. The basic gemstone chromophores are copper, iron, manganese, chromium, nickel, cobalt, vanadium and titanium. The mineral beryl, beryllium aluminum silicate, is an example of an allochromatic gemstone; with iron present, the color becomes blue or blue-green and the stone is aquamarine. When chromium is present, the color is a deep green and the stone becomes the much more valuable emerald.

Early Hindus, noting great similarity in hardness and weight between the corundum gemstones, believed that sapphires were really "unripened" rubies. Today, we know that sapphires and rubies are allochromatic, for pure crystalline corundum is colorless. The red of rubies is created by a trace of chromium oxide incorporated into the crystal structure. Every other color of corundum, including pink, yellow, green, blue and violet, is considered as sapphire. The blues, the classic and most valuable color of sapphire, are caused by trace amounts of iron oxide or titanium oxide, or both.

The visual appearance of any gem also depends upon the manner in which it reflects and refracts light. Light reflecting from the surface of a stone creates luster. In transparent stones, only a very small part of the light is reflected from the surface. The transmitted light is refracted, or bent, within the crystal and reflected from internal surfaces back toward the source to create the effect of brilliance. Generally, the higher the index of refraction, the greater the brilliance. Diamond has an extremely high index of refraction—2.4; that of sapphires and rubies is 1.8, considerably higher than many other gems.

Refraction also causes fascinating color effects. If sufficiently refracted, white light will separate into its spectral components, a phenomenon known as dispersion. Within a diamond, dispersion creates the tiny flashes of spectral color known as "fire." In a colored stone, such as ruby or sapphire, the dispersion effect is seen within the range of the predominant color. A well-cut sapphire will therefore exhibit its brilliance in flashes of color in a wide range of blue and even violet.

Some gems, including most sapphires, are doubly-refractive, creating yet another color effect. The beam of light striking the stone may be split into two beams at opposing angles, each of which may be absorbed differently. When shifted ninety degrees during viewing, such stones will change color. Such dichroic sapphires may appear blue when viewed from one direction, and blue-violet from another. Dichroism is, therefore, an important consideration in cutting.

"Perfect" mineral crystals or gemstones are theoretical, for microscopic examination virtually always reveals objects that are foreign to the basic crystal composition and structure. These inclusions fall into four general classes: solid particles of a different mineral; cracks or other openings in the crystal that have been subsequently filled by foreign material; cavities or "bubbles" containing gases, liquids or small crystals; and uneven distribution of color, or color zonation. All gemstones contain inclusions that vary greatly in type, number and size, and which often indicate origin of the stone. Large or highly visible inclusions detract from both the beauty and value of a gem. Some inclusions, however, may actually enhance appearance. In sapphire and ruby, tiny, needle-like inclusions of rutile, titanium dioxide, may align themselves in directions related to the crystal structure and reflect light internally. In the hexagonal corundum gemstones, these reflections may form three crossing bands of light—the six-pointed "star" effect of star sapphires and star rubies.

Ancient cultures were fascinated with gems, even though they were unable to bring out the full beauty that we know and admire today. Their crude techniques of shaping, cutting and polishing were not able to maximize the effects of the light that struck their gems. While many softer stones could be easily worked, sapphire, ruby and diamond, because of extreme hardness, were only roughly shaped and poorly polished. The first true artificial faceting, in which man attempted to duplicate the facets found on natural mineral crystals, was not performed until the Middle Ages. These cut gems had only six or eight facets, but they served their intended purpose of allowing more light to enter the stone to increase color and brilliance. For centuries thereafter, European guilds closely guarded the secrets of gem cutting. It was not until the early nineteenth century that cutting equipment was improved and cutting techniques became common knowledge. Today, gem cutting is based on the science of optics which dictates the proper angles and proportions that will maximize brilliance.

In prehistory, the richest source of sapphires and rubies were doubtlessly the alluvial "gem gravels" of Burma and Ceylon. These stones supplied the early gem traders, became the basis of the gem tales in *Sinbad the Sailor*, and were described in the writings of Marco Polo. The Burmese deposits were probably the first to be systematically mined. Early Burmese miners dug simple, shallow shafts to reach the *byon*, or layer of gem-bearing gravel, then hauled the gravel to the surface for washing and recovery of the gemstones, primarily rubies. The first modern mention of Burmese gem mines came in 1597 when the King of Burma took them over from a local ruling Shan. For centuries, the deposits were worked intermittently, sometimes with slave labor. By the late 1870s, oppression, smuggling and thievery had created such chaotic conditions as to make mining impossible. About 1880, the Burmese King, Thebaw, leased mining rights to the British-controlled Burmah and Bombay Trading Company, but arbitrarily cancelled the lease in 1882. This action, along with other provocations, prompted the British to invade Burma in 1886 with an army of 30,000 men. The British annexed Upper Burma to the colony of India, then promulgated the Upper Burma Ruby Regulations, creating the Mogok Stone Tract from the historic gemstone deposits.

In 1889, the British government awarded control of the Mogok mines to Edwin Streeter, the London jeweler who organized Burma Ruby Mines, Ltd. In the years that followed, the British moved the entire town of Mogok, built roads, bridges, washing plants and even a hydroelectric plant for power. Their daily problems included heavy rainfall, isolation and jungle diseases. Another were the hundreds of hired native workers who would stop at nothing to pilfer the rough sapphires and rubies. In the sorting plants, British managers required the natives to wear cardboard boxes with gauze windows over their heads to prevent them from swallowing the gemstones. The British employed high powered water jets to erode away the alluvial banks, forcing the mud through a series of sluice boxes for gemstone recovery, a method known as hydraulicking. The mass mining proved successful and, by late 1895, the Mogok stone tract was a major supplier of rubies and sapphires to the European gem markets. Only a year after Edwin Streeter secured the mining rights to the Mogok Stone Tract, he would journey to Montana on behalf of the Sapphire and Ruby Mining Company of Montana to be duped in the fraudulent mining venture on the Missouri River bars.

Like those of Burma, the gem deposits of Ceylon have been legendary for centuries. In 334 B.C., Nearchus described an island not far from Persia where beautiful translucent gems were mined—doubtlessly the island of Ceylon. In the early 1500s, Portuguese sailors began trading for the Ceylonese gemstones which were mined from the gravels, locally called *illam*, by primitive methods similar to those used in Burma. Among a broad variety of gemstones, the most important was the blue sapphire. The Ceylonese mine production, the result of thousands of individual native miners, has always been a major source of the world's sapphires.

Gem mining in Siam was first reported in 1850. By 1880, two mining districts made the nation another important producer of sapphires and rubies. Again, it was the British, led by Edwin Streeter, who gained control of important Siamese gem deposits in 1895. Mining was conducted under the auspices of Sapphires and Rubies of Siam, Ltd.

One of the most legendary gemstones of all time, the Kashmir sapphire, was discovered in 1881 as a result of a landslide on the slopes above a glacial cirque at the elevation of 14,000 feet. The location was one of the most rugged and remote on earth—the Zanskar Range in the Himalayan Mountains of northwest India, described in early reports as "the region beyond the snows." Incredibly, local villagers first traded the large sapphires, called *nilam* ("blue stone"), for salt on a weight for weight basis. Within a year, the quality of the large blue sapphires had greatly excited gem dealers. In 1883, the maharajah of Kashmir took control of the deposit. For the next five years, Army officers controlled all mining. The stones produced at that time were of such quality and color that the term Kashmir would come to signify the most desirable and expensive of blue sapphires. After that, political problems, the small size of the deposit and severe climatic conditions combined to keep production sporadic at best.

In the 1870s, another source of sapphire was discovered in Queensland, Australia. These deposits produced blue sapphires that were very dark and tinted with green, and that turned nearly black in artificial light. While the Australian blues fell short of the best oriental blues, the greens and yellows were considered among the world's best.

In 1896, when Jim Ettien and Jake Hoover were staking claims to the Yogo dike in central Montana, the world sapphire market was supplied primarily by the oriental mines. The name Kashmir had already become one of great prestige, but the main sources of sapphires, as they had been for centuries, were Burma and Ceylon. So closely associated was the image of the Orient and sapphires, that green sapphires were commonly referred to as "oriental emerald," yellow sapphires as "oriental topaz," and violet sapphires as "oriental amethyst."

The new discoveries and increased mine production of the middle and late nineteenth century meant a greater supply of sapphires than at any time in history. Still, the growing demand for fine sapphires in jewelry, then the vogue in Europe, continued to outpace the supply. The British enjoyed at least partial control over every major sapphire source and market and were aggressively expanding their interests. When George A. Wells, resident of Great Falls, Montana, and representing the New Mine Sapphire Syndicate, walked into the London offices of Johnson, Walker and Tolhurst, Ltd., the British gem merchants had been eager to extend their influence to that new deposit called Yogo.

As the nineteenth century drew to an end, a new sapphire, distinctive in quality and color, was about to make its appearance on the world precious gem markets, a sapphire that came not from the Orient, but from the United States of America.

Chapter 3

The English Mine

Unlike Montanans in the distant, remote Judith River Basin, Johnson, Walker and Tolhurst, Ltd., from their London offices, enjoyed a commanding view of the world sapphire markets. From the beginning, they knew they had a potential bonanza in the Yogo dike. Moving quickly, they had needed only two years to gain control of the New Mine Sapphire Syndicate, and only two more to realize complete ownership. Under the early American ownership and management of the Syndicate, the Yogo sapphire had made few advances, either in production or marketing. But when British capital, together with engineering and marketing expertise, came into play in 1898, things changed quickly. Mine production increased sharply and the world began learning about Yogo sapphire. In 1900, Yogo sapphires were exhibited in the Universal Exposition in Paris. In direct competition with oriental sapphires, they were awarded a silver medal.

In 1900, mine production had again amounted to 400,000 carats, including 125,000 of gem quality, with a total value of $80,000. Since several years of hydraulicking had depleted the most easily accessible dike rock, plans were drawn for an underground mining opera-

The certificate accompanying the silver medal awarded to Yogo sapphires at the 1900 Paris Universal Exposition.

tion. Developments at the Yogo dike had already caught the attention of much of Montana and were receiving increasing mention in the press. On November 2, 1900, Yogo sapphires were on the front page of the *Great Falls Tribune*.

> **TO SINK DEEPER IN THE YOGO SAPPHIRE MINES THIS WINTER**
>
> A SHAFT AND LEVELS TO BE RUN TO DETERMINE WHAT DEPTH THE FORMATION EXTENDS—THE LEAD FIVE MILES LONG
>
> The Yogo sapphires in Fergus County are to be developed through the winter

> and in the spring a hoist will probably be installed for the purpose of tracing the deposit to depth. This information came from State Senator S. S. Hobson, manager of the New Mine Sapphire Syndicate, which owns the mines.
>
> "We have stopped washing for the season and shipped the last clean-up to London after making an output which exceeds all previous seasons. But we shall continue mining all winter. The highest points are worked by open cuts, but we had sunk a shaft 100 feet and next year shall run levels and put on a hoist to sink deeper. The lead has been prospected for nearly five miles in length, but we would like to sink, for curiosity as well as to open ground ahead. The character of the ground seems to be the same at depth as near the surface. There are places where it is harder than others, but so far we have been able to work with hand drills and it slacks with exposure to the air. The government geologist thinks it goes clear down.
>
> "The stones lately have been better than ever in size and our London agent reports a better demand than ever for them. We received a handsome medal at the Paris Exposition for their fine quality and, so far as I know, we were the only firm in Montana that received an award. Our exhibit put in the shade that from Ceylon, which has hitherto held the palm. In fact, I have learned that dealers in precious stones have been in Helena trying to sell Ceylon sapphires as coming from Yogo."

Hobson's last comment was quite true. The sudden availability of a high-quality American sapphire in the classic blues which had previously come only from the Orient was fostering considerable dealer misrepresentation. Yogo sapphires quickly became the object of home state pride but, since all mine production was now shipped to London with few cut stones yet returned to the United States, few Yogos were

Early surface workings at the Yogo dike. The timbers, or stulls, were used for ground support and to prevent loose rock from falling on the miners below.

A Yogo miner stands alongside the "sinking bucket" in the first shaft ever sunk into the dike. In this photograph, taken about 1899, the shaft is only about ten feet deep. Eventually, the English Mine underground workings would reach a depth of over 250 feet. Although a steam hoist would be installed as the workings progressed deeper, the English Mine never brought in modern mining equipment.

available locally. To capitalize on the local demand for a Montana blue sapphire, some enterprising jewelers simply purchased Ceylonese sapphire, proclaimed them to be Yogos, and reaped the profits.

While Hobson was aware of oriental stones being passed off as Yogos locally, he was not aware of the European marketing deceptions where just the opposite was taking place. Many European gem collectors had been conditioned to believe that the only sapphires worth acquiring were those from the traditional and "exotic" sources of Burma, Ceylon, Siam and the Kashmir. In the world of sapphires, the name "Montana," or more specifically "Yogo," neither impressed nor appealed to potential retail purchasers. European jewelers began purchasing the Yogos for their highly marketable beauty, color and quality. But, since disclosing their true origin might mean losing a sale or compromising on the highest possible price, some misrepresented the Yogos as oriental sapphires. Such misrepresentation usually befell the bigger Yogos. Since Yogos were decidedly smaller than their oriental counterparts, they picked up an enduring reputation as "small and flat," not all of which was deserved.

Misrepresentation of origin was done only at the retail level. Johnson, Walker and Tolhurst, Ltd., as the Yogo sapphire jewelry wholesaler, went to considerable lengths to promote the stones as American. By 1900, a dazzling line of "New Mine" sapphire jewelry was on the market, including sapphire and pearl enameled necklaces, sapphire and diamond brooches, and sapphire throat ornaments and pendants, all set in gold. In their first New Mine sapphire jewelry catalogue, Johnson, Walker and Tolhurst printed a lengthy statement by Claremont and Ward, a leading London gem cutting firm:

> We have recently cut from the rough upwards of 11,900 carats of these beautiful "New Mine" sapphires, and we have, during the progress of our work, taken the greatest pains to compare them, in every way possible, with those sapphires which come to us from Burma, Siam, Ceylon, Australia and elsewhere.

As gem cutters, Claremont and Ward found the Yogos to be a refreshing change from the usual run of alluvial sapphires coming from the Orient and Australia. Since they were mined from a rare lode deposit, they had not been subjected to alluvial wear. Retention of their original crystalline shapes was proving an advantage to the cutters.

> It is, however, comparatively rarely we meet with a crystal of sapphire perfect in shape, as these gems come into the market generally in broken or water-worn fragments, which show upon examination just a characteristic indication of the system to which they belong. This fact is due to the thousand and one vicissitudes incumbent upon such a precarious existence as gems are subject to, and from the moment of their formation to their appearance in Hatton Garden or some other gem cutting centre as marketable products.
>
> Now a very noticeable feature of the "New Mine" sapphires is that they are more often found in the perfect geometric form designed by nature than are the sapphires from other parts of the world. True it is, however, that the figures even of crystals of sapphire which reach us from the other hemisphere in a perfect crystallographic condition are often of such a complex nature that the task of deciphering the relation-

> ship of one face with another requires a crystallographer of no mean ability, and we have nothing but admiration for the vocabulary equal to adequately describing the wonderful and beautiful surface marking of these precious stones in the rough—protruding and receding angles, parallel striae and flanges, and a dozen other interesting effects, all of which are intimately connected with the hexagonal system.

Claremont and Ward then commented on other unusual features of the Yogo sapphire, namely the absence of inclusions, cloudiness, and color zonation, which greatly facilitated the job of cutting the rough gemstones.

> All kinds of gems, as everybody knows, sometimes contain flaws, feathers, and other imperfections, the successful removal of which depends upon the skill and judgment of the lapidary. The most difficult imperfection with which the lapidary had to contend in sapphires is the presence of cloudy or semi-opaque patches within the stone, often occurring in parallel lines which generally form a series of hexagons or triangles one within the other. This defect often mars the beauty of a costly gem, and by its removal immense loss of weight is incurred, not infrequently accompanied by deterioration of color.
>
> The "New Mine" sapphires are absolutely free from this tiresome effect of cloudyness, which is undoubtedly the reason for the great brilliance and luster.
>
> With regard to colour of these gems—they range from the palest steel color through all the different shades of blue until they reach, in the finest specimens, that lovely tone called "cornflower blue," which, until comparatively recently, was associated only with the sapphires of Burma and Siam.
>
> Moreover—and this is particularly striking to a gem cutter—the colour is always quite evenly distributed throughout the stone, and never found in patches as is the case with all other sapphires; therefore the "New Mine" sapphires, when cut and polished, cannot possibly appear "parti-colored."

Just as the color, quality and beauty of the Yogo sapphires were attracting the attention of jewelers and gemologists, the source of the sapphire was of great interest to geologists who were intrigued not only with the configuration of the dike, but by the nature of the dike rock itself. At the surface, the twenty-five million-year-old solidified magma had been altered by the elements into a yellow-gray, granular, crumbly clay. At a depth of about forty feet, however, unaltered dike rock was encountered. This was a green-gray friable rock, solid enough to require light blasting for removal. After mining and subsequent exposure to the elements, even the unaltered dike rock weathered and deteriorated rapidly. Mechanical crushing, which would have destroyed many sapphires, would never be needed. After mining, the dike rock deteriorated within a few months to a crumbly clay ready for washing. Visiting geologists most often compared the Yogo dike rock with the "blue earth" of the South African diamond fields.

Interestingly, a number of other similar igneous formations were found in the immediate vicinity of the Yogo dike, the largest only 900 feet to the north and paralleling the dike for one mile. This formation, too, had been thoroughly prospected, but not a single sapphire was ever found. Only the Yogo dike contained sapphires, for reasons still not understood.

The Yogo sapphire had quickly become the nation's most unusual mineral product. Interest remained high because of the steadily increasing profits and the unusual methods of mining. Accordingly, the *Great Falls Tribune* ran another front-page Yogo feature on June 3, 1901.

THE YOGO SAPPHIRE MINES

WHERE ARE EXTRACTED FROM THE SOIL THE GEMS FOR WHICH LAPIDARIES OF THE UNITED STATES, GREAT BRITAIN AND FRANCE EAGERLY COMPETE

Vein Traced For At Least Five Miles At Great Depth

Process Employed In The Mines Very Similar To Those Used In Mining For Gold

The sapphire mines of the Yogo mining district are now recognized as being among the most valuable gem mines in the world, and the stones produced from them rivaling in radiant chatoyancy the famed sapphires of Ceylon, and second in value to the diamond only, have become

One of the many weathering heaps at the English Mine. Note the trestle at upper right. The trestles were quickly disassembled and moved to new locations as needed to spread the ore over a greater area for maximum exposure to the elements.

equally desirable and almost as popular as an article of personal adornment as the latter gem. The great vein in which the magma holding gems lies is in itself a most interesting and attractive study for the geologists. . . .

By far the greater part of the workings of the New Mine syndicate have been executed upon the surface and much of it has been done with the hydraulic and by ground sluicing. The company conveys its supply of water through a ditch and flume about ten miles in length, from Yogo Creek, and a complete sluicing system extends over the entire length of their present workings, in addition to which there is a pipeline for hydraulic work, reaching to all parts of the vein undergoing exploitation.

The oldest workings in these mines, which were begun in 1896, are located

The weathering heaps. Newly mined ore in tiny ore cars was pushed along rails atop the trestles, then dumped into the weathering heaps. Note that the "floor" is timbered to minimize the sapphire loss. In several months, the large pieces of dike rock in these heaps will have deteriorated into a clay-like material suitable for washing.

about one mile east of the summit of Yogo Hill, and there are several great open cuts on the vein, one of which is 700 feet in length, and has an extreme depth of 90 feet. Just east of this cut is the "Blue Diamond" cut, having a length of 1,200 feet, and varying in depth from 20 to 50 feet, with a width of from 8 to 20 feet. In addition to these there are several other large cuts, and a tunnel in one place has been driven 600 feet, having a depth at the face of at least 200 feet.

The vein at the surface, and sometimes to a depth of fifty feet, is much decomposed, and the hydraulic force is sufficient to work it for that depth, but in the tunnel just mentioned the vein-stuff has attained a much greater degree of hardness and the ordinary system of mining in vogue in quartz mines is used in the extraction. . . .

The method of extracting the sapphires from the magma is much similar to that used by placer miners in sluicing for gold. The sluice boxes are fitted with Hungarian riffles, as in placer mining, but it is necessary to exercise more care in setting them, as the specific gravity of the sapphire is much less than that of gold, and with the boxes set at too great a "pitch," they would pass over the riffles and be lost. The magma is first passed over a "grizzly" (sizing grate), and the finer parts pass at once into the sluices. The coarser, harder material is thrown into the dump, where, after exposure to the elements, it finally disintegrates and is then passed into the sluices. The process of disintegration occupies from one month to a year according to the hardness and tenacity of the magma. . . .

For cleaning up an ordinary rocker is used, with three screens with as many sizes, through which the "pay dirt" is passed. When the "pay dirt" has been worked down as closely as possible in the rocker it is panned in one of the finer screens, dried and the stones picked by hand from the small amount of gravel remaining. All the stones are saved, the larger perfect ones being valuable as gems; the smaller ones being used as watch jewels and in other fine mechanical work; the imperfect ones are ground and mixed with diamond dust and used in gem cutting and for polishing purposes, when it is known as emery.

The stones from the Yogo vein vary greatly in size from the smallest to four or five carats, and as gems they are equal in value to the best products of the mines of Ceylon and Burmah. The largest stone yet found in the New Mine Syndicate mines weighed a little more than nine carats. . . .

Workmen cleaning a sluice box. The fence-like device behind the workmen are metal riffles which have been removed from the bottom of the sluice boxes. The riffles act as a trap to retain the heavy sapphires during the washing process. Sapphire recovery at the English Mine was very similar to placer gold recovery methods.

Production at Yogo had risen steadily from the 400,000 carats of 1900 to one million carats in 1903. The percentage of total production classified as gem material varied greatly within different sections of the dike, ranging from a low of only five percent to as much as one-third. The bulk of production, unsuitable because of imperfections, shape or size for cutting into gems, was classified either as industrial sapphire or as industrial corundum. The industrial sapphires, while too small or flat for gems, could be cut into bearing jewels for fine instruments. The corundum was suitable only for manufacture into abrasive compounds. As mechanical manufacturing of all types was booming in the early 1900s, a strong demand for industrial sapphire had made even the non-gem Yogo production quite valuable. George Frederick Kunz offered a fascinating description of the watch jewel industry published in the United States Geological Survey's 1903 *Mineral Resources*.

At no former period were watch jewels made so beautifully perfect as to mechanical accuracy. A certain number of jewels, often simply called stones, are used in every watch. A watch is said to run on so many stones, and though it cannot strictly be said that the value of a watch increases with the number of stones used, still in an approximate sense it is true. This is indicated by the fact that during the last fifteen years, which have witnessed a very marked improvement in watches, the number of stones required for the works of a first-class watch has been increased by nine, and as millions of watches are made annually, the number of jewels sold annually is at least from 10,000,000 to 20,000,000. The little gems are pierced to receive the gearing of the axles of the wheels. The object of using them is to give to the works a base which shall cause the least friction and shall not wear out easily. Among the gems used for this purpose, garnet is the least valuable, but it is much used in the cheaper watches. Sapphire and rubies, fine enough in quality to make gems, are mostly used, but only minute pieces are necessary. For the most part, however, these gems are merely fragments of the large ones which have no color, or else are rolled crystals that are such color as to have no value, and hence are not considered as jewels. This is especially true of sapphires too pale for setting, which, however, are a shade harder and hence more serviceable for watch stones, and of stones which, like the Fergus County, Mont. (Yogo), blue flat crystals, or the Granite County, Mont. (Rock Creek), multicolored crystals, have little value in jewelry. Many thousands of ounces of these American gems are sold at from $1 to $5 per ounce, and are an important factor in American sapphire mining.

In Switzerland most of the jewels are cut and sold in boxes of from 500 to 1,000 per box. Each stone has been given a rounded form and is pierced in the center, the drill-hole being smaller by a minute quantity than the diameter of the axle it is to hold. The bed of the stone in the watch is a small cylinder, apparently of brass, but in reality consisting of a soft-gold alloy. Before the stone is handed to the watchmaker it is put in a lathe, and by means of a tiny steel drill, covered with oil and diamond dust, the central opening is enlarged to enable the steel axle of the pin for which it is intended to fit into it accurately. The watchmaker first fixes the cylinder in the lathe, then picks up the stone with the moistened finger and inserts it into the cylinder while the latter is turning with the axis of the lathe. With a pointed tool the workman next presses against the edge of the revolving cylinder and thus forces the soft metal to cover and protect the sapphire or ruby to such an extent that it appears as if imbedded in a metallic cushion. Next a drill is inserted in the metallic coat of the cylinder from the opposite side of the lathe, and a hole is drilled in this coat exactly of the same size as the hole in the stone itself. A great variety of forms have been made recently, not only for watches but for electric and other meters. The latter, as compared with watches, require a greater and more enduring

> life in the jewels, which, owing to the microscopic inclusions, either of softer minerals or fluid cavities, is often shortened materially. Sapphires, rubies and even diamonds are used with wonderful ingenuity, and with increasing demand for hard bearings in the endless variety of electrical devices, in which the moving parts revolve rapidly, there is much to be looked for in the way of new devices, and a greatly increased demand for jeweled bearings is probable.

Yogo sapphires were already quietly finding their way into millions of the world's finest watches, but the gem quality stones were what earned both publicity and profit for the New Mine Sapphire Syndicate. The 1903 mine production of one million carats had an overall value of over $100,000. The bulk of production, 7,000 ounces (450 pounds) of industrial sapphire and corundum, was worth about $25,000. The 40,000 carats (266 ounces, or about sixteen pounds) of gem sapphires were valued at $80,000 in the rough. Since operating expenses that year were about $60,000, the New Mine Sapphire Syndicate was turning a very substantial profit.

A far greater profit was being made in London, after Johnson, Walker and Tolhurst cut and polished the rough Yogos into fine gems. Cutting retention, the percent of weight retained in the finished gems, was about 35 pecent. All that remained of the 1903 mine production of 40,000 carats of gem quality rough was about 14,000 carats of cut gems. But those now had a value of $20 to $25 per carat, making the 1903 mine gem production—after cutting—worth well over a quarter-million dollars.

If the $20-per-carat value of the cut sapphires does not seem "precious," remember that inflation has greatly eroded the value of the dollar. In 1900, only $15 would buy a new Winchester Model 94 rifle and $18 would be enough for a first-class rail ticket in a sleeper car from Montana to New York.

The directors of the New Mine Sapphire Syndicate, most of whom were connected with Johnson, Walker and Tolhurst, Ltd., found Yogo a convenient, exclusive source of quality sapphires and a very profitable one at that. There was never any intention of putting the profits back into modernizing and mechanizing the English Mine. The directors were concerned only that operating expenses be held to a minimum while maintaining the supply of rough gem sapphires. How to dispose of the profits was a simple matter; by 1902, the directors of the New Mine Sapphire Syndicate began declaring hefty

Yogo sapphire jewelry was pictured in a 1904 catalogue published by Johnson Walker & Tolhurst, Ltd. For such elaborate designs, the uniformity of color of the Yogos proved a great advantage to jewelry manufacturers.

dividends that would rise to as much as 40 percent annually.

In 1899, T. Hamilton Walker and his assistant mining engineer had been accompanied by a third Englishman, Charles T.Gadsden. The twenty-five-year-old Gadsden had neither education or experience that particularly qualified him for work at a gemstone mine. He was apparently in need of a job; being offered one, he took it, even if it meant living and working in a place called Yogo, Montana. Under T. Hamilton Walker, Gadsden served as a general assistant, performing

A formal photographic portrait of Charles Gadsden taken at a Great Falls studio. The year was 1902 and Gadsden was about to return to England to marry. He would return in April, 1903, with his wife, Maude, as supervisor of the English Mine.

whatever odd jobs were necessary. Over a period of three years, he demonstrated some managerial ability, competence in practical matters, integrity, and, most importantly, an unquestioned company loyalty. In 1902, the Syndicate directors acted to minimize operating expenses and maximize dividends. Among their moves was the recall of T. Hamilton Walker and his engineering assistant. In a somewhat surprising move, Charles T. Gadsden was named resident supervisor of the English Mine. Gadsden was not destined to be a passing figure or play a bit role in the Yogo story. Instead, Gadsden would become the very essence and soul of the English Mine. For the next half century, Yogo would be the private little empire of Charles T. Gadsden.

The surface erosion that finally exposed the eastern two-thirds of the Yogo dike had done so gently and evenly. At the western end, however, the topography was much more dramatic, beginning where Yogo Creek had slashed deeply through both the Madison limestone and the contained dike. Further west, the dike had weathered away first into a trench, then into a watercourse flowing east into Yogo Creek. In time, the water had eroded away both the dike and the adjacent limestone to form a deep canyon the early settlers knew as Kelly Coulee.

When Jake Hoover was staking his lode claims to the dike in 1896, he doubtlessly realized that the dike extended westward beyond the bench lands and into the rugged limestone cliffs of Yogo Gulch and even into Kelly Coulee. Since this area was poorly suited for surface mining of the dike, Hoover and his partners did not stake claims west of the bench lands. The local sapphire excitement of 1896 did attract a number of other curious prospectors to the Yogo mining district, among them two employees of a nearby sawmill, John Burke and Patrick Sweeney. Burke and Sweeney traced the dike west into Yogo Gulch and Kelly Coulee, filing six lode claims on July 4, 1896, and naming their property the "Fourth of July Claim." Two years later, encouraged by developments at the nearby English Mine, Burke and Sweeney decided to begin sapphire mining. Digging out weathered dike rock was a simple matter, but washing was not. Lower Yogo Creek, especially in late summer, carried little water, a problem compounded by the long ditch and flume system that diverted water from the upper creek to the English Mine. The ore had to be hauled four miles in wagons to the Judith River for washing. Mining was slow and expensive, and Burke and Sweeney sold the relatively few rough sapphires they recovered locally, never further away than Great Falls, Billings, Lewistown or Helena, and never achieving a profit.

Although John Burke and Pat Sweeney would never make a penny mining sapphires, their claims were becoming quite valuable, thanks to the considerable publicity focused on Yogo sapphires and the already successful English Mine. In 1901, a newly formed New York City company, the American Gem Syndicate, stepped forward with an enticing offer. The company would lease the claims from Burke and Sweeney, provide them with management positions at the mine, and aggressively market the sapphires in the United States. Burke and Sweeney took the offer, only to quickly learn the American Gem Syndicate was unable to raise any operating capital. Further encouraged by the profitability of the English Mine, Burke and Sweeney decided in 1903 to do the job themselves. Their operation, already known as the "American Mine," to distinguish it from the English Mine, received its first publicity on March 14 in the *Great Falls Tribune*.

MONTANA SAPPHIRES TO BE CUT AT MINES

Sapphires from the Burke & Sweeney mines, at Yogo, are to be cut at the mines by expert lapidaries, and hereafter Montana sapphires will be placed on the New York and other markets ready for sale as soon as received. In the past, sapphires from Montana have been sent to New York, Paris or London, and were cut in those cities, but Burke & Sweeney have decided that better results may be secured by having the gems cut at the mines, and when they are placed on sale, they will be a Montana product in entirety.

Expert lapidaries, who have been engaged for several years in the cutting of Montana sapphires, have been engaged by Burke & Sweeney to do the cutting at the mines. Several thousand dollars

> worth of machinery for just this purpose has been ordered, and just as soon as weather conditions shall become such that the necessary water power may be secured, the cutting of the gems at the mines shall be commenced.
>
> John Burke, who, with P. T. Sweeney, owns the properties, was in the city yesterday, making arrangements for the marketing of the finished products of the mines. There has always been a greater demand for Montana sapphires than could be met, and there is no doubt that the entire output of the Burke & Sweeney properties will be in great demand, as the stones are of very superior quality. The owners are not concerned as to securing a market, but are seeking one in which they may sell to the greatest advantage.
>
> "We can not tell at present," said Mr. Burke, "how many men we will employ. That will depend upon the demand for the stones, and we expect that to be large. At present we are employing but a few men, but the force will be greatly increased as soon as we are able to obtain water, of which we have a great supply in season.
>
> "We have been operating the properties since the middle of last month. For a time prior to January 12, they were tied up by a bond, held by the American Gem Syndicate, which would not proceed with development. The bond having expired, we took over the properties and expect to work them to a great extent and market the cut stones as rapidly as they may be wanted."

The optimistic report meant nothing; John Burke and Pat Sweeney, like many would-be sapphire miners who would follow them, were naive about the difficulties of mining the dike and marketing the sapphires. Four months of back-breaking work yielded only a few thousand carats of gem material which didn't begin to cover expenses. Burke and Sweeney had failed, but still managed to walk away from the Yogo dike with more money than either of them had ever seen. In 1904, they struck it rich when the American Sapphire Company paid $100,000 for the outright purchase of their "Fourth of July" property. This new company, like the old American Gem Company, was based in New York City, but was somewhat better funded. Even though only a fraction of the authorized capital of $500,000 had been raised, the American Sapphire Company plunged into development of the American Mine. Choosing a site on Yogo Creek where a large vertical cross section of the dike was exposed in the limestone cliffs, the company immediately drove a short tunnel eastward into the dike. Unlike the open bench lands of the English Mine, the cramped creek bed of Yogo Gulch did not provide sufficient room for weathering dumps. To attempt to accelerate the weathering process, a $30,000, 100-ton-per-day mill was constructed on Yogo Creek.

By 1907, mine development had been completed and the American Sapphire Company was theoretically prepared to begin production. But funds were now exhausted and the company was still desperately trying to sell out its initial stock offering in the face of a general economic depression, which both minimized available risk capital and curtailed purchases of luxury items, such as gems and jewelry. To make matters worse, the gems cut by lapidaries in Great Falls and Helena were judged inferior to European-cut stones. Facing adverse publicity, and unable to sell its stock or sapphires, the American Sapphire Company declared bankruptcy in 1909.

The last steam cloud over the mill in Yogo Gulch had hardly dissipated when a newly formed Montana firm, the Yogo American Sapphire Company, headquartered in Great Falls, purchased the assets of the American Mine for $60,000. Like its predecessors, this new American company had used all its funds to acquire the American mine; its first order of business was not directed to solving the problems of mining and marketing, but to selling investment shares to raise operating capital.

As the English Mine rolled along on the eastern section of the Yogo dike and the various owners of the American Mine struggled on the western end, other sapphire mines were sporadically active in western Montana. The Missouri River bars, where Montana sapphires were first discovered, were turning into a graveyard of failed corporations. The original venture, the Sapphire and Ruby Mining Company

The American Mine mill in Yogo Gulch. Built in 1905 for a cost of $30,000, the mill was able to treat about 100 tons per day. It was last used in 1913 and was finally torn down in the early 1950s. The cleft in the limestone cliffs is the Yogo dike. Natural water erosion had worn away the dike rock leaving only the walls of Madison limestone in place.

of Montana, which was exposed as a fraud in 1894, was reorganized in 1897 as the Eldorado Gold and Gem Company with the Spratt Brothers, still the claim owners, as chief officers. Simultaneously, two smaller companies, the Spokane Sapphire Company and the Montana Gold and Gem Mining Company, both probably inspired by developments at Yogo, also went to work on the bars. All failed within a few years for the same reason—the inability to market the varicolored sapphires. The green and green-yellow stones were never in demand, and the pink and yellow sapphires, although brilliant and quite beautiful, were neither plentiful nor valuable enough for a profit. The blue sapphires that might challenge the Yogos were extremely rare.

Commercial sapphire mining at Rock Creek, southwest of Philipsburg, did not begin until 1899, seven years after the discovery. Two seasons of hydraulicking yielded a half million carats, 25,000 of which were gem quality. The gemstones, mostly pink and pale red, were cut in Helena. The operation was not profitable and the claims lay idle until 1905 when they were acquired by the American Gem Mining Syndicate of St. Louis, which owned a Swiss watch jewel manufacturing company. In 1911, production even exceeded that at Yogo, both in gem and industrial sapphires, yet the American Gem Mining Syndicate only broke even, while the English Mine racked up another solid profit. The difference, as always, was the minimal gem value of the vari-colored sapphires.

The Dry Cottonwood Creek sapphire deposit, discovered in 1899, was not commercially mined until 1902 when the Northwest Sapphire Company, of Butte, established a short-lived hydraulicking operation. The deposit was again mined in 1907 when the Variegated Sapphire Company brought in a large, floating mechanical gold dredge. The dredge was dismantled the next year; in 1910, the Consolidated Gold and Sapphire Mining Company, also of Butte, tried their luck with a smaller dredge. That dredge was no more successful than its larger predecessor, never recovering more than a few thousand carats per year, and never coming close to a profit.

Many Montana miners were learning the basics of sapphire mining and marketing the hard way. It was quickly demonstrated that industrial sapphire and corundum could not be mined for their own sake. Montana had countless millions of carats of vari-colored gem sapphires, but already a string of broken companies had proven that they were neither particularly valuable, marketable or profitable. Success clearly rested in top quality gem sapphires in the classic blue colors that could rival, and often surpass, the sapphires of the Orient. And there was only one deposit on the North American continent where such sapphires could be mined—the Yogo dike.

When Charles Gadsden accepted the position of resident supervisor of the English Mine in January, 1903, he immediately began preparing for what he somehow knew would be a lifetime at Yogo. Visiting his old home near Berkhampstead, he quickly married an acquaintance of many years, Maude Margaret Mellor, the daughter of a chemist who owned a London drugstore. At the time, Maude was attending a prestigious Berkhampstead finishing school where the alumnae would include such luminaries as the future Lady Winston Churchill. In Charles Gadsden, Maude had chosen not only a husband, but a different life than that for which she had been prepared.

On a cold, blustery day in early April, 1903, Mr. and Mrs. Charles T. Gadsden stepped down from a stagecoach onto the dusty main street of Utica, Montana. Behind them were handed down their two large steamer trunks, still plastered with the stickers of the White Star Line. The last twelve miles of that long journey from England were completed in a lurching, horse drawn wagon. At the mine, Maude found her new home to be a cramped, two-room apartment on the second floor of the company office building. Maude was a quiet woman, but one of great inner strength, for she bridged the wide gap between life in the upper middle class suburbs of London to that in rugged, turn-of-the-century central Montana with little apparent difficulty.

Even in his first year of authority, Gadsden confirmed the wisdom of the New Mine Sapphire Syndicate directors in naming him supervisor. Gadsden kept a sharp eye on every aspect of the operation and came up with many innovative ideas which, in the absence of any substantial development capital from London, always allowed him to

The town of Utica as it appeared about 1900. For many years, the New Mine Sapphire Syndicate's shipments of Yogo sapphires were loaded onto stagecoaches on this street to begin their long journey to London, England.
Utica Historical Society

maintain production. Gadsden recruited most of his miners from the booming hardrock copper camp of Butte, paying them the standard wage of three dollars per day. Most of the mines in Montana and the West were undergoing mechanical modernization, but the English Mine was not among them. Instead of preparing blasting holes with pneumatic drills, Gadsden's miners relied on the ancient technique of hand drilling, manually pounding chisel-like hand steels with four-pound sledges. And rather than electric haulage and lights, mules still pulled tiny rail-mounted ore cars while the miners used candles and carbide lamps. Visiting American mining engineers would shake their heads at Gadsden's underground mining operation, describing it as "primitive" and "backward." Still, the English Mine profited steadily, while many capital-intensive American mines failed.

Gadsden devoted a great deal of attention to treatment of the ore after it was mined. At first, the ore was simply spread out on the ground for maximum exposure to the elements. As it deteriorated and was moved about, the sapphires began working their way lower until some became lost in the ground. Suspecting such a loss, Gadsden constructed large weathering "floors," consisting of heavy planks on timber foundations which could be swept clean, thus assuring that virtually every sapphire in the ore reached the sluices. To maximize the recovery efficiency, Gadsden designed and built a series of four dams along a lightly declining gradient. After the initial washing, the tailings, or "slums," filled the pond behind the highest dam. When the pond was full, the slums were passed through sluices into the next pond, then into the next, always through the sluices. The system assured the deterioration of even the smallest bits of ore, and thus a very high recovery efficiency.

Another matter demanding Gadsden's attention was that of security, for the gem quality rough sapphires, on a weight-for-weight basis, were considerably more valuable even than gold. A three-carat rough sapphire was only the size of a bean, yet it could be worth twenty dollars, roughly a miner's wage for an entire week of drilling, blasting and shoveling. It was an understandable human failing for a miner to be tempted to slip a sapphire into his pocket when no one was looking. This threat of "highgrading" was not unique to the English Mine; owners of most western gold mines had been plagued with the problem for decades. Although clearly a form of theft and illegal in every court, miners considered highgrading an earned right, paid for

A group of miners posing in front of the shaft house of the English Mine. Charles Gadsden recruited many of his miners from Butte. Not all were pleased with the isolation and somewhat lower wages of the English Mine.

in the physical risk and drudgery of underground mining.

Without a doubt, Yogo sapphires disappeared into miners' pockets, but Gadsden apparently minimized the loss. He personally oversaw cleanup of the sluices four times daily. His hired men were allowed to tend the sluices, clearing them of gravel as necessary, but Gadsden always enforced the old rule of the western placer mining camps: *keep your fingers out of the riffles*. When Gadsden caught, or had good reason to suspect a highgrading miner, he paid the man's "time," escorted him to the mine gate and left him to walk the twelve miles to Utica.

As Yogo became a household name in central Montana, there were more and more requests for tours of the mine. Gadsden obliged the curious whenever he could, but even considered his most distinguished guests to be potential highgraders. Even men of the cloth proved they were not above the temptation to pocket a Yogo sapphire. The *Great Falls Tribune* once told of a local preacher who, upon finishing his tour, thanked Gadsden profusely for his courtesy—only to be shocked by the supervisor's request that he empty his pockets. Along with the usual keys and coins came two rough sapphires. As

Miners at the English Mine at work in the very early 1900s. The two miners on the right are using an out-dated hand drilling technique known as double-jacking; the miner on the left holds the drill steel while his partner swings away with an eight-pound sledge. The holes were then loaded with light charges of dynamite to blast the solid dike rock loose.

Gadsden solemnly reclaimed the stones, the red-faced preacher tried to laugh it off as a "joke." Out of respect for the ministry, Gadsden did not prosecute. But justice, of sorts, was done when the newspapers gave the incident front page attention. One writer commented, "I believe I would rather hide my shame in jail than be publicly proclaimed as a sapphire thief."

There was, literally, a fortune in sapphires lying not only in the sluices, but in the weathering dumps. During one tour, a Great Falls high school science teacher was intrigued by the great stream of water that played over the weathering dumps. Falling behind his group, he wandered over to the dumps and sorted through the deteriorating ore. Finding a "large, bright pebble," the teacher asked, "By chance might this not be a sapphire?" Gadsden gave the piece a cursory examination, then used his pocketknife to pry the "pebble" free of its matrix. Placing the stone in his pocket, he replied in his

Part of the weathering heaps and system of trestles at the English Mine. This photograph, probably taken about 1906, preceded the installation of timber weathering floors.

clipped British accent, "By chance that might indeed be a sapphire. Perhaps three carats. Our best in some time. Thank you."

One newspaper writer brought his niece, a Texas schoolteacher, to tour the mine. His later account of that visit made humorous mention of the omnipresent temptation among nearly all visitors to claim a "souvenir."

> . . . We were fortunate for they were making the annual clean-up. At the sluice boxes was a pile of little sapphires, a wagon load it seemed, such as are used for bearings and watches and instruments of precision. Then Mr. Gadsden held a double handful of sapphires, the size of pigeon eggs, for our inspection. Emily's eyes sparkled, and I said, "Watch Emily's fingers, watch her fingers!"

Charles Gadsden also demonstrated a practical mechanical ingenuity that found many applications. Among the heavy minerals that accompanied the sapphires in the sluice concentrates were large amounts of common pyrite, fragments of tiny crystals of iron sulfide called "fool's gold." Where the British engineers had been unable to

The English Mine was a maze of wooden trestles, tramways, flumes and sluices. In this photograph, workmen are reducing the volume of sluice concentrate in a wooden rocker, a recovery operation very similar to that employed in placer gold mining.

devise a simple method to separate the worthless pyrite, Gadsden succeeded. Placing the concentrates in a large metal pan, he applied gentle heat. The heat both dried the concentrate and made the pyrite magnetic, permitting easy extraction with a homemade electromagnet.

Gadsden salvaged parts from an old washing machine to build a classifier for the sapphires. He rigged two rubber rollers in the configuration of a narrow "V" with the closed end slightly elevated, then attached the whole thing to an old bicycle frame. By pumping the pedals he was able to rotate the rollers, keeping his hands free to feed a stream of rough sapphires onto the rollers. As the stones worked down the rollers, they dropped through the increasing width of the opening, falling into receptacles below in order of size.

While the mine supervisor was a model of ability and efficiency, Charles Gadsden, the man, was somewhat of an enigma. Even though the lanky Englishman had no love for the United States, both he and Maude quickly became naturalized citizens. Gadsden regularly sent lengthy letters to the Syndicate directors in London, detailing every aspect of the mine operation, often digressing into caustic criticisms of everything American, including the political system, economy, government and the American people themselves. One individual Gadsden particularly disliked was Jake Hoover, whom he would never speak with unless another man were present as a witness, plainly considering the discoverer of the Yogo sapphires a liar. Fortunately for both, their contact was infrequent for, in 1903, Hoover was about to head north to pursue his golden dreams in Alaska.

While Charles Gadsden made no true friends in Montana, nor seemed to want any, he was not without compassion. The nearest settlement to the English Mine was the ruins of Yogo City where old Millie Ringgold still lived. Crippled by age and rheumatism, Millie had left Yogo City in 1903 to accept county relief. But after a month in a Great Falls home, Millie so ached to return to her home of twenty-five years in the Little Belt Mountains that she was permitted to leave. For the last years of her life, Gadsden ordered the mine wagon and team to bring Millie her supplies. In December, 1906, Millie was found gravely ill on the floor of her cabin. Dr. Abram Poska was summoned from Utica and remained with her until she died two days later. When the English mine wagon and team made its last trip to old Yogo City to bring out the body of Millie Ringgold, Charles Gadsden himself was handling the reins. Although her last wish was to be buried in Yogo City, the old ex-slave was laid to rest in Utica. In the simple, makeshift way of the frontier, her casket was an old piano box.

Millie Ringgold was an exception, for Gadsden rarely allowed himself to be distracted from mine operations and problems, one of which was tailings disposal. Since only a few carats of sapphires were extracted from every ton of dike rock, Gadsden had to dispose of a voluminous quantity of tailings. In frontier mining, disposal was usually by the cheapest and most convenient method. During the early years of the English Mine, the tailings, or "slums," were simply dumped at the eastern end of the mine property where waste water from the sluices eventually washed much of it into the Judith River. As they were swept downstream, they began silting irrigation ditches of farms and ranches. Believing the slums might be detrimental to agriculture, the farmers and ranchers banded together and filed suit in Fergus County Court, demanding that Gadsden be prohibited from

Old English Mine workings. The overhead stulls provide ground support. This section was hydraulicly mined and the ruins of the sluice boxes can still be seen.

further dumping of slums into the Judith. The court granted an injunction and sapphire production in 1906 and 1907 was dramatically curtailed.

The answer, from an engineering standpoint, was construction of more dams to create tailings and settling ponds that would retain the slums while releasing only clean water into the Judith, a considerable expense that the Syndicate directors would not be eager to finance. Gadsden, as usual, found an inexpensive solution. The prize vegetables that he and Maude harvested from their gardens grew in nothing but slums. Not only weren't the slums detrimental to agriculture, they actually seemed quite beneficial. Gadsden explained this to the directors, requesting and receiving the necessary funds to purchase the 560-acre Pagel Ranch that adjoined the eastern end of the mine property. He constructed a simple sand trap to remove any larger pieces of the tailings, then directed the fine slums over the ranch land. With the injunction still pending in spring, 1907, Gadsden planted oats, alfalfa and vegetables. At season's end, the "slum farm" had produced, in the words of a visiting government geologist, "the most luxuriant crops, probably, in the Judith Basin." With the "slum farm" as evidence, the injunction was lifted and the mine returned to full production next spring. The slums, with their high content of available nitrogen and phosphorus, were a very effective natural fertilizer.

Gadsden even built a local reputation as a successful livestock man, raising horses and mules for draft animals, and cattle, goats and sheep for meat, both for his own table and those in his miners' mess hall. After 1907, much of his feed came from regular alfalfa harvests on the "slum farm." Another of Gadsden's many interests was breeding and using carrier pigeons. The lifeline of the mine was the twelve-mile-long ditch and flume system that diverted water from upper Yogo Creek. Gadsden's men carried a pigeon on their frequent inspections of the ditch. If a break or weak point was observed, a quick note was scribbled and inserted into the pigeon's leg capsule. Minutes later, Gadsden had received the message and could take appropriate action.

The *Great Falls Tribune* reported that Gadsden's flock of twenty homing pigeons was also employed far beyond the English Mine.

> Whenever Mr. Gadsden leaves the mine for some other point in the state, he usually carries with him one or more of the pigeons to send back communications. About ten weeks ago he came to Great Falls at which time he brought with him three pigeons. All three were released with messages during his stay here, two of them getting through in a couple of hours but one appears to have become lost and did not return to the mine until two months later.
>
> It is said that when Gadsden communicates with some other mining man or other personage in Butte (Butte was 120 miles from Yogo) or some other point in the state and desires an immediate reply he sends this person one of his pigeons, by means of which he gets his communications within two or three hours from the time when it is written.

Gadsden's broad interests and abilities, of course, were all directed toward maintaining a steady supply of Yogo sapphires, which he kept personal control of until they were shipped. Maude would stitch the stones in rows between two layers of heavy muslin. Each piece of cloth was then sewn tight, rolled up, placed in a canvas bag, sealed and sent to London by registered mail. Never once was there a problem or loss in shipping.

A section of the Yogo dike showing the old workings of early English Mine days. The walls that form the sides of the workings are 300 million-year-old Madison limestone.

A typical sapphire shipment. Maude Gadsden fashioned this striped pouch. Inside was a tobacco can containing a sewn muslin fabric holding the sapphires. All sapphires were shipped to London by registered mail. There was never one problem in shipping during the history of the English Mine. This particular shipment, postmarked 1924, was among the last. The return address is "From C. H. Gadsden, Utica, Montana."

By 1910, the English Mine was composed of a group of unpainted, rough bunkhouses, a general store, school and shops clustered about the largest structure, which was the enclosed headframe, or, in British terminology, the "shaft house." Surrounding the little community was an extensive maze of wooden trestles and tramways radiating outward from the shaft house to the weathering dumps. Beneath the elevated tramways wound another network of flumes to convey water to wherever it was needed. Although the English Mine appeared bleak and depressing, it certainly did its job. In 1910, production topped a half million carats of industrial and watch jewel material, as well as 100,000 carats of gem sapphire. The 1910 production, a record, was valued at $200,000.

The period around 1910 brought the recovery of some of the largest Yogo sapphires on record. Even the struggling American Mine, still under the control of the Yogo American Sapphire Company, seemed

The English Mine shaft house, or head frame, and boiler room. Most of the sapphire production of the English Mine was hoisted to the surface through this building.

to be having a bit of luck and received some badly needed publicity in the *Great Falls Tribune* on September 8, 1910.

> **BIG SAPPHIRE FOUND AT YOGO**
> WEIGHS NEARLY SIX CARATS AND IS SAID TO BE THE FINEST EVER FOUND IN AMERICA
>
> What is said to be the finest sapphire ever found in Montana, which means in America, or anywhere in the world for that matter except in India, is now on exhibition at Sutter Bros'. jewelry store, at Lewistown. . . .
>
> The stone weighs 5¾ carats, of a beautiful deep blue color, with brilliant luster, rivaling the diamond and being fully as valuable as a diamond of that size. It is a perfectly faultless stone and well worth the seeing.
>
> It has been pronounced the best stone

> by Mr. (George A.) Wells of the English company, who has just returned to England after a visit to the Yogo mine. It was found in the property of the Yogo American Sapphire Company, in Fergus County, and is another evidence of the supremacy of this county.

The well-publicized American Mine sapphire was soon overshadowed by two stones coming from the English Mine. One was twelve carats, the other a nineteen-carat stone the size of a peach pit, the largest sapphire ever to come out of the Yogo dike. Neither of the English mine sapphires received much publicity at all; Maude dutifully sewed both into the muslin layers and quietly shipped them to London.

Most of the Yogo sapphires reaching London were cut, manufactured into jewelry, and marketed in Europe. At first, few were returned to the United States, probably because of the high tariff imposed on the entry of foreign-cut gems. Gadsden took it upon himself to be the leading promoter of the "New Mine" sapphires in the United States. He created the name "Montana Blue," and continuously pressured the directors for promotional literature to distribute. Some jewelry stores, mostly in Montana and the Northwest, set up window displays to promote the "Montana Blue" which included many photographs of the mine, all courtesy of Charles Gadsden. Among Gadsden's better advertising efforts was the distribution of "Montana Blue" literature on the Great Northern Railroad's passenger trains. Gadsden's motivations to expand the Yogo sapphire market in the United States had little to do with increasing profits, for those, as shareholders' dividends, only went into the directors' pockets. More likely, Gadsden sought to increase the demand for Yogo sapphires to assure continued operation of the mine which had now become his own little private empire.

Gadsden sometimes went to unusual lengths to assure that nothing on the property escaped his attention. In 1913, the Syndicate agreed to provide the supervisor and his wife with more suitable living accommodations. Gadsden himself chose the location and designed the house. The location was atop a low hill about 150 yards south of the dike that afforded a commanding view of the mine. The house itself was L-shaped so Gadsden would have an unbroken field of view of all of the surface workings. Gadsden even busied himself keeping abreast of business at the American Mine. When the first telephones were installed, they were all tied to the single line from Utica. The American Mine telephone was the end of the line which conveniently passed not only through the English Mine offices, but also through Charles Gadsden's living room. In his correspondence to London, Gadsden mentioned with apparent pride that he listened in on American mine business "as much as is possible when I have the time."

By 1913, the booming gem business in Europe had at last convinced the directors of the wisdom of investing a part of their dividends into expansion of mine production. On July 13, 1913, the *Great Falls Tribune* ran a front page feature on the new plans, sapphire marketing and the general state of the colored gem industry.

BIG IMPROVEMENTS AT SAPPHIRE MINE

LONDON CAPITALISTS ARRIVE IN CITY ON WAY TO YOGO TO DIRECT WORK OF EXPANSION OF NEW MINE SYNDICATE'S PROPERTY

GREAT DEMAND FOR MONTANA BLUE STONES, THEY SAY, IN EUROPEAN MARKETS—PRICES SOARING

Large improvements are to be made at the sapphire mines at Yogo, the object being the greatest possible output for the mines at the earliest possible date. The expansion will be started at once and will be pushed under the personal direction of officials of the owning company, the New Mine Sapphire Syndicate, three representatives of the company arriving in this city yesterday direct from London to take up the work.

The men in the party are Vice President F. H. Wood and Secretary Sidney Finnegan of the company and F. H. Lathbury, an expert mining engineer who will determine the plans for the enlargement and expansion.

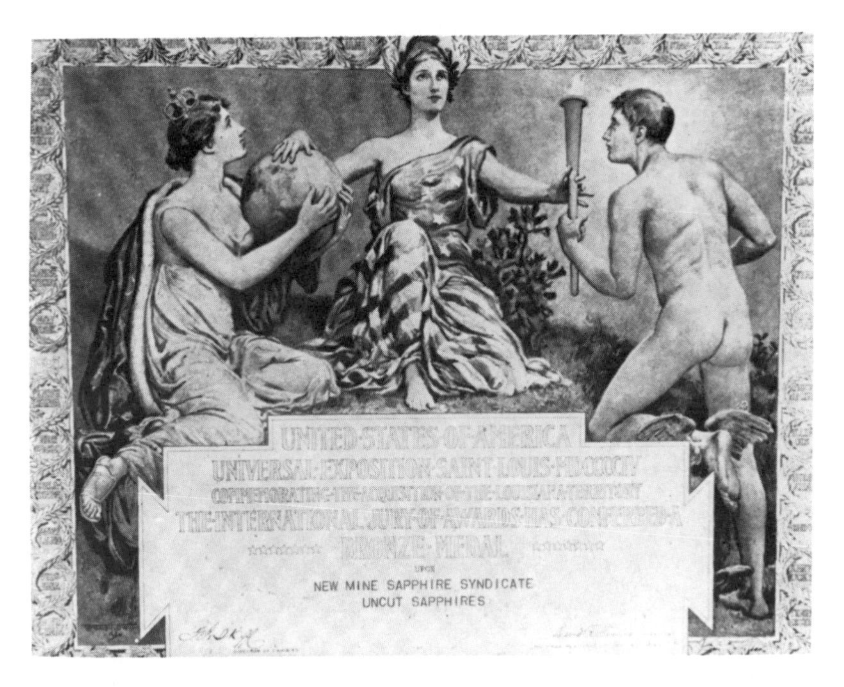

The certificate accompanying the bronze medal awarded to the New Mine Sapphire Syndicate for uncut Yogo sapphires at the 1904 St. Louis Universal Exposition.

"There has been a rapidly growing demand for the Montana sapphire," said Mr. Wood, "and recently the demand has been so heavy that we have decided to do what we could to meet it with our product and incidentally reap some of the profit that it will mean. Sapphires have been gaining in popularity and with that, of course, has been an attendant increase in value. Just to impress what the increased demand has brought with it in price, let me cite a particular case. One London jeweler recently had in his possession a five-carat Montana sapphire from our mine. He sold it for 55 British pounds sterling per carat, making it bring 275 British pounds sterling. That, approximately, was $1,375 for the sapphire. But

This promotional display for "The Blue Montana"–the Yogo sapphire–was placed in the window of a prominent jewelry store in Seattle, Washington, shortly after World War I. The display included many photographs of the English Mine and refers to the Yogo as "the United States' Most Important Gem.".

the man to whom he sold it resold it just a few weeks later for 90 British Pounds the carat or for approximately $2,250 for the sapphire.

"There has been a very substantial reason for this tendency toward the sapphires by people seeking to purchase gems. The prices of pearls and diamonds have soared to such a point that carrying them in large quantities means a heavy risk for the jeweler should new fields be discovered of such size as possibly to force the prices down. Under such a condition high-class jewelers endeavor to turn their customers' attention to other desirable but less expensive colored gems and in that line they could have found nothing more beautiful than the Montana sapphire. After all,the sapphire is much like the blue diamond and the ones who like gems are coming more and more to realize what elegant gems the sapphires are.

"We have been working with 35 to 40 men at the mine and we shall hope to increase that force as rapidly as possible. We sunk the mine an additional 100 feet last year, as you probably know, and we are now working at the 250-foot level. We are getting the finest gems we have ever secured and in abundance and we expect very soon to be turning out an output of very materially increased proportions. . . ."

B. P. McNair, who transacts considerable of the company's business at this end of the line, was given a very pleasant surprise by Mr. Wood's party yesterday. They brought as a gift to him a handsome gold ring with a very perfect sapphire as the setting and presented it to him at his office yesterday afternoon. It makes a very valuable present.

The planned expansion of the English Mine never took place, for the available capital would soon be diverted to another end—acquisition of the American Mine. The Yogo American Sapphire Company, like the three previous owners of the American Mine, had yet to make a profit. The long line of problems had included technical mining difficulties, insufficient room for ore weathering, a frequent lack of water, the poor quality of work turned out by United States cutters, and the perennial inability to market Yogo sapphires in the United States. At the end of 1913, the Yogo American Sapphire Company, of Great Falls, which had struggled for four years, declared bankruptcy and put the property up for sale.

Charles and Maude Gadsden journeyed to London in the spring of 1914 to persuade the Syndicate directors to acquire the American Mine. Knowing their supply of rough sapphires was certainly adequate, the directors argued. Gadsden countered on logic; the property was available now for a bargain price and its acquisition would eliminate any possible future market competition. Gadsden, of course, was doubly motivated by his own personal reasons. The presence of another mine on "his" dike had always irritated Gadsden, and to gain control of the entire dike would be to extend the bounds of his little empire. Charles Gadsden managed to convince the directors. When he and Maude returned to Montana two months later, he carried with him the funds to purchase the American Mine. In May, 1914, the New Mine Sapphire Syndicate acquired all the assets of the Yogo American Sapphire Company, and thus ownership and control of the entire known length of the Yogo dike.

During the summer of 1914, Charles Gadsden devoted his full attention and manpower to the workings of the newly acquired American Mine. Knowing that there had been no room at the base of Yogo Gulch to properly weather the ore, Gadsden surmised that sapphire recovery must have been grossly inefficient. To prove his theory, he ordered his miners to rewash the old tailings that lay heaped about in Yogo Gulch. When ice halted the rewashing operation in October, Gadsden had recovered rough sapphires worth $80,000—exactly enough to offset the purchase of the American Mine.

Meanwhile, in London, Johnson, Walker and Tolhurst, Ltd., was preparing a new promotional campaign for the New Mine sapphires centered around a comprehensive, fifty-page booklet titled *A Royal Gem–A Monograph of the Sapphire, with a History and Description of the New Mine*. This was probably the first jewelry industry effort to establish a specific identity for a gem. Previously, gems had been loosely associated with their origins, such as Burmese rubies, Colombian emeralds, Ceylonese sapphires, etc. But now the New Mine sapphire was being presented as a stone of special origin and special qualities that clearly distinguished it from any other sapphire. The booklet was an innovative and original combination of blatant commercial advertising and constructive public education about sapphires. After a somewhat elaborate, Victorian opening, the booklet listed the qualities of the New Mine sapphire and challenged those of the Orient.

> The beautiful word sapphire is a synonym for blue; to use it with an adjective qualifying colour is an entymological crime. Sapphire! the very sound conjures up visions of tropical seas, of deep placid lakes, of

A general view of the English Mine about 1914. The large building is the shaft house which is surrounded by trestles, weathering heaps and tailings dams.

the azure dome of a cloudless sky, and of the modest gentian clustered many-headed in a sunny pocket on the mountain side. No sapphire exists which is not blue; there are few shades of blue which the sapphire cannot match.

From the deep black of a moonless night—almost black, but for a mysterious gleam of indigo—through the finest gradations in colour, to the perfection colour of the cornflower, thence to the most delicate of steely blues, the sapphire—more particularly the "New Mine" sapphire—delights the eye of the lover of beauty. Truly, it is a Royal Gem.

The jeweller will probably find the sapphire represented in his stock in far greater numbers than all gems, except the diamond. This is because the public favor the sapphire, and because the steady supply during the last decade of evenly coloured and well matched sapphires from the "New Mine," has provided the industry with a very effective and dependable "line."

The sapphire is unsurpassed as an associate to the bold brilliance of the diamond. It represents a strong colour-contrast without harshness; by co-operation it magnifies the blue and violet rays, and by negative action modifies and almost overcomes any traces of yellowness the diamond may possess. These useful properties are possessed by sapphires to a far greater extent than by other colored gems. The ruby will emphasize the impure colour of a brownish diamond; an emerald will magnify a yellow tinge in a by-water stone; but it is always safe to use a sapphire.

Until comparatively recent years the world's supply of sapphires came from the East. India, Ceylon, Siam and Burma produced these gems in various quantities and markedly different in colour. The Kashmir mine, situated in the Zanskar Range of the north-western Himalayas, had the credit of producing gems more nearly approaching "perfection colour" than other mines. The occurrence of sapphires in the ruby mines of Burma is comparatively rare. Ceylon sapphires are distinguished by a most disturbing patchiness of colour. Indeed, the colour is so severely localised in many crystals that none but the most expert lapidaries are able to "make" gems of any size with success. A great quantity of good sapphires come from Siam, and the mines in Anakie, Queensland, produce a grade of very dark, greenish sapphire which is not sufficiently attractive in colour to achieve popularity as a gem.

The gems produced at the "New Mine" are sapphires of rare distinction and quality. They have an evenness of colour, an almost complete absence of "silk," and milkiness, are favoured with a freedom from blemishes and impurities, and match up in a very practical and useful manner.

One most remarkable attribute, and a feature adding a great artistic and commercial value to the gems, is their radiance when viewed by artificial light. Oriental sapphires, excepting only the very finest and most expensive, have the one great failing of absorbing artificial light and of appearing black and lustreless under these circumstances. The "New Mine" sapphires reflect artificial light as well as daylight, and, as a consequence, when worn at night with the lights coming from many directions, the "New Mine" sapphire gains in life and beauty, giving ray for ray and glow for glow in merry duel of brilliancy with every flashing light.

In the effort to educate the retail jeweler and the public about sapphire origin and quality, *A Royal Gem* also touched upon gem cutting which, for centuries, had been a closely guarded secret held by guilds. To further enlighten the reader, the cutting and polishing processes in use at the time were described.

Cutting is the process of shaping up the gem and grinding facets on the surfaces. Polishing is the final operation of removing the roughness left by faceting and imparting the fine surface seen upon a finished gem.

The lapidary selects a crystal and after a critical examination decides how and where it shall be divided for cutting. Very rarely is a crystal of any size cut without a preliminary slitting into two or more pieces. The cleavage of the sapphire is very distinct, but the practice of slitting the gems on the edge of a small rapidly rotating metal disc, fed with diamond dust and oil, is a surer method of division. After slitting, the portion of the sapphire to be cut is cemented to a wooden holder; the stone is now applied to the surface of a revolving lap of soft metal charged with diamond dust, each portion being presented to the action of the lap in turn until the necessary facets are applied.

The holder is held in the lapidary's hand with its upper end steadied against a small vertical post having a series of notches or holes at varying heights, corresponding with the angles at which the holder is usually placed. Mechanical aids for the correct formation of facets are now in use, but the more skillful lapidaries use their own judgement and depend on this for correct shaping and angles of the facets. After the upper part, or crown, of the gem is cut, it is unset and turned over in the cement. The base or pavilion facets are applied in a similar way.

To polish a sapphire the lapidary uses a similar holder, but a wooden lap covered with leather and a softer abrasive. The gem is cemented to the holder and each facet is carefully held to the face of the lap until it loses all marks and abrasions of the cutting process and acquires a high polish. The gems are then cleaned, sorted into parcels of a convenient size and character and are ready for the market.

By the end of 1914, the New Mine Sapphire Syndicate seemed to be in a very strong position. It owned the entire Yogo dike, enjoyed record European sapphire prices, and saw Yogo sapphire demand increasing as Johnson, Walker and Tolhurst's promotion took effect. Gadsden continued to urge the directors to step up production. This time, however, his argument fell on deaf ears for in London, war, not sapphires, had become the primary concern. Charles Gadsden had already nearly become an early casualty of the war. When he and Maude journeyed to England in spring, 1914, they sailed both ways on the White Star liner *Lusitania*. Less than a year later, a German submarine torpedoed and sunk the *Lusitania*, an event that would help drag the United States into the growing war. The Syndicate directors understandably foresaw the war as a serious and lengthy disruption of the European gem market. In a decision they would later regret, the directors ordered Gadsden to forget any ideas of expanding production. Instead, Gadsden was to slash his operating expenses by cutting operations back to a maintenance level. In 1915, Gadsden laid off over half his work force, suspended mining, and assigned his remaining men to the rewashing of old tailings. For the remainder of World War I, mine production at Yogo was minimal.

After the Allies had successfully concluded the war, the Syndicate directors made plans to step up production. Francis H. Wood, then the president of the New Mine Sapphire Syndicate, visited Montana to discuss the matter with Gadsden. Wood's comments were reported in the *Great Falls Tribune* on July 4, 1920.

MONTANA SAPPHIRES NOW FINDING READY MARKET IN MANY FOREIGN LANDS; COMING TWO YEAR'S PRODUCT OF YOGO MINES IS ALREADY CONTRACTED FOR

Montana sapphires, the finest in the world, produced from the Yogo mines at the Yogo properties of the New Mine Sapphire Syndicate in southwestern Fergus County, have enjoyed such an increase in popularity in recent years that the demand, formerly existing only in America, has extended to Europe with the result that the syndicate now has requisitions upon its books amounting to a half million dollars consuming the entire output of the mines for the next two years.

Such was the information given out here last week by Francis H. Wood, of London, England, president of the syndicate upon his return from a stay of a month at the workings.

"Within the past three or four years the selling value of the Montana sapphire has doubled, in fact the very finest of them has trebled in value," said Mr. Wood. "The demand is increasing as the stone gets better known and we are now selling it in the rough in South America.

"We are also marketing it in India, which is a triumph as India is the home of the oriental sapphire.

"Before the war we had made all preparations to sell it in Italy. We are also opening a market in China. As far as England is concerned, there has always been a prejudice against the Montana sapphire but by means of propaganda, advertising to the trade and the interest of leading jewelers, we have removed this prejudice and now I do not believe one can look in the window of a good jewelry store in London without finding there a Montana sapphire. When I came away from London we had requisitions on our books amounting to a half million dollars.

"The last time I was out here, in June, 1913, I told of our intentions to expend $20,000 in the development of our mines. This money was spent for that purpose, but the next year war was declared. We thought the luxury trade would suffer and we cabled Manager Charles T. Gadsden at the mines to close down. . . .

"But we found that instead of money being close, the working classes and middle classes had more money than ever. About fifteen months later, we saw a big demand coming and we then tried to resume operations. We were getting along very well until America came into the war whereupon the miners all quit and got into the essential industries.

"Since that time, we have never been able to be fully under operation. Labor was scarce and when, last year, we made a good start the water gave out, and as we depend on water the same as in placer mining, we were unable to wash out gems, although the lack of water did not prevent us from getting dirt out of the mines. Last August the water gave out to such an extent that there was not enough available even for a pot of tea.

"This year we have gotten a force of men at work and are now employing about fourteen men underground and while we have a force on the washing crew we still need men. We are now paying miners $4.25 per day and board while the surface men are drawing $4.00 per day and board. . . ."

At the present time work is being carried out on two levels from one shaft only. At the 100-foot level the mine has been worked to the extent of 3,000 feet, while from the 200-foot level 1,800 feet have been penetrated. This limestone formation or dyke has a depth of 1,200 feet, with an average width of eight feet, and as it extends for a distance of about five miles, it can be seen that many years will be required to work it out. . . .

To the casual reader, Yogo seemed troubled by a minor lack or water from the two previous dry summers and a bothersome labor shortage. The future, if one were to believe Francis Wood, seemed bright indeed and was secured by high sapphire prices and booming worldwide demand for Yogo sapphires. In truth, however, the Syndicate was now facing the gravest wave of problems in its history. Because of Gadsden's shoestring budget, the English Mine was one of the lowest paying mines in Montana. Even though the Syndicate had grudgingly raised a miner's wage from $3.00 to $4.25 per day, a good miner could make better money in the Butte copper mines. The post-war industrial boom was also luring many men away from the mines. Gadsden found his labor costs rising and the quality of his miners declining.

Still, production in 1921, thanks to mass washing of ore mined during the previous three years, equaled the recovered production of 1911. But while mine production was encouraging, marketing was suffering in a major post-war change of jewelry styles. The smaller sapphires, which Yogo could produce in enormous quantity, were no longer the vogue. Large sapphires, of which there were few, were in demand. With an inventory overloaded with small sapphires, the Syndicate was forced to shift its marketing effort from Europe to the United States.

Another monumental problem was the matter of post-war taxation. As a British-owned corporation registered and operating in Montana, the New Mine Sapphire Syndicate became liable to a barrage of taxation. United States and British corporate income taxes amounted to 22.5 percent *each*. Added to that was a Montana state corporate tax of 5.5 percent. The New Mine Sapphire Syndicate was, therefore, faced with a staggering tax rate on profits of 50.5 percent. Yet another financial concern was the British pound sterling. Once one of the world's strongest currencies, the pound was becoming a victim of Britain's war-strained economy and inflation, and was losing ground relative to the dollar. In 1900, dollar operating costs at Yogo, relative to the value of the pound, were minimal. Now, increasing costs coupled with a declining exchange rate multiplied the cost of mining Yogo sapphires substantially.

As if these problems were not enough, the Syndicate had growing concerns over the long-term stability of the colored precious gem markets caused by the appearance of synthetic gems. Synthetic rubies were created by a flame fusion process in France as early as 1866, when they also caused a minor panic in the jewelry establishment. The flame fusion process was greatly advanced by Auguste Verneuil, a French chemist, in the late 1800s. The Verneuil Process, as it became known, was simple but at first difficult to control. Highly pure, powdered aluminum oxide was dropped through a stream of oxygen and into a very hot gas flame. While descending, the tiny bits of aluminum oxide melt, then fall onto the tip of a revolving rod where they crystallize. The rod continues to increase in length as it is lowered. The result is a single, pencil-shaped corundum crystal called a boule. With the addition of a trace of chromium oxide, the boule became red—synthetic ruby. Verneuil could "grow" ruby boules weighing fifteen carats in only two and one-half hours. At first, the synthetic Verneuil rubies were a curiosity at gem expositions. But when the details of Verneuil's techniques were published, many European laboratories rushed to produce synthetic rubies. By 1908, production had reached five million carats annually and created a panic among ruby buyers worldwide. Among the mines where production was dramatically curtailed was the Burma Ruby Mines, Ltd., operation on Burma's Mogok Stone Tract, the source of the finest rubies in the world. Burma Ruby Mines, Ltd., would never fully recover from the turmoil and uncertainty that struck the ruby markets in 1908.

Satisfied with the quality of his synthetic rubies, Auguste Verneuil then turned his attention to sapphire. At the time it was not known which element or elements were the chromophores responsible for the blue color of sapphire. With a bit of experimentation, Verneuil found them to be iron and titanium. Addition of trace amounts of iron oxide and titanium oxide to the powdered aluminum oxide produced sapphire boules of beautiful blue colors. This process was published in 1911; by 1913, laboratories were producing six million carats of synthetic sapphires annually. World War I halted both production and research in synthetics but, by 1920, the laboratories were again working to further refine the processes and, thus, the quality of the synthetic sapphires. This research was understandably of great concern to the New Mine Sapphire Syndicate. Most jewelers could not distinguish the synthetics from natural sapphires; only a gem expert could detect and recognize the subtle "growth rings" and swirls of tiny, microscopic gas bubbles that characterized the synthetics. If improved techniques could produce synthetics *without* those telltale signs, the bottom could well fall out of the natural sapphire market. The immediate threat would not be to the oriental and Australian sapphires with their obviously natural array of faults such as off-colors, color zonation and inclusions. The most vulnerable stones would be the most perfect sapphire that nature had ever created—the Yogos, with the perfect blues and virtual absence of inclusions and color zonation.

Historically, about one-quarter of the value of English Mine production came from non-gem material—bearing jewel and abrasive corundum. Synthetic corundum, of course, had the identical hardness as natural corundum. Bearing jewels and abrasives of the future, clearly, would come not from mines, but from laboratories. In the early 1920s, the value of natural sapphires—and of sapphire mining properties—in the not-too-distant future was a matter of some conjecture.

One thing was clear; not another penny was going into improvement of the English Mine. In 1922, Charles Gadsden managed only to expand the area of his wooden weathering floors to a quarter-million square feet—equivalent to a square 500 feet long on each side. In 1920, even as Francis Wood was telling the *Great Falls Tribune* of the Syndicate's bright future, the directors had probably already agreed that it was time to pull out of the Yogo dike. The Syndicate had profited enormously for over twenty years. The only years a dividend had not been declared was 1902. The average dividend for every other year had been a hefty fifteen percent, a rate very few other investments could have matched. But now the time had come to cash in the assets and get out. In 1922, Charles Gadsden and a local attorney drafted a prospectus for the sale of the Yogo property. After listing earnings and dividends as proof of the mine's potential, the prospectus pointed out other possible values of the property and the terms.

> In addition to the property's value as a sapphire mine, it is also valuable in three other respects:
> (1) There is good evidence that gold could be profitably mined along Yogo Creek on sites discovered since placer mining was abandoned in 1898. However, geological studies indicate that no large deposits approaching a mother-lode could be present.
> (2) The property, because of its scenic beauty, accessibility by auto, rich farmland, water rights and proximity to Lewis & Clark National Forest lying along its southwest border, is ideal vacationing and hunting. A dude ranch and hunting camp is profitably operated within three miles of the mine property.
> (3) Excellent crops have been grown on the tillable land.
>
> TERMS OF SALE: This Syndicate, a Montana Corporation, is offered for sale for $150,000, this price to include all land, water rights, machinery and buildings owned by the corporation. Since it would be necessary to secure approval of the present stockholders before sale could be actually completed, 10% of the purchase price would be required in cash before meeting of the stockholders would be called. This meeting would be held in Montana. Mr. Charles T. Gadsden, the present mine manager and a large stockholder, controls all necessary proxies to approve the sale. The remaining 90% of the purchase price would be payable in cash within six months, or in accordance with any agreement approved by Mr. Gadsden. The majority stockholders, being English citizens, have considered the problem of exchange and require payment only in dollars.

Events at Yogo were reflecting the declining fortunes of British territorial and business interests worldwide. The sprawling British Empire and realm of influence were crumbling and beginning a defensive withdrawal back to Britain. In the Orient, the Burma Ruby Mines, Ltd., was undergoing liquidation and other British gem mining interests were in their last years. As the Syndicate made plans to sell the English Mine, Charles Gadsden was making plans for his own future, and they involved no travel. Whatever sale might be arranged, one of the terms would be that Charles Gadsden would remain as resident supervisor. Gadsden had already spent twenty-five years at Yogo, and he had no intention of leaving now.

The proposed sale of the mine became only a hypothetical issue for, in 1923, an event took place that determined the ownership and management of the property for the next three decades. On July 26, a hot and humid Thursday, Gadsden tended his sluice boxes, watching the slow but steady buildup of towering thunderheads in the western sky over the Little Belt Mountains. By late afternoon, the sky had turned black and the air, heavy with humidity, crackled with static. At half past four the rains started and continued as a deluge for two and one-half hours. The rainfall, although not measured locally, may have been as high as four or five inches. The gentle hills of Madison limestone, unable to absorb the cloudburst, collected the water and channeled it in torrents down through the workings of the English Mine. At dusk the skies cleared and Charles Gadsden stared down from his hilltop home at what the waters had done to his little empire. Dams were broken, weathering floors, ore and sapphires had been swept away. Tramways had been undermined and collapsed, and sluices and flumes had been broken up and washed as far down as the Judith River.

In the days that followed, Gadsden sorted through the wreckage, estimating the cost in time, money and effort that would be needed to repair the damage and put the mine back into production. He then sat down at his desk, and dutifully penned a letter to Syndicate president Francis Wood, who was then in Seattle personally attempting to market the high inventory of small Yogo sapphires.

July 28, 1923

Dear Mr. Wood:
. . . On Thursday 26 we had about 2½ hours rain which will entirely spoil our season output, it has taken out Dam No. 2, raised all the lower end of floor and it must be rebuilt, both floor and Dam. It swept away all the flume below Dam No. 2 and washed the iron bottom on to Dam No. 4. Dam 3 & 4 also gone.

Half of the good dirt at the old camp has gone. The logs which went at Dam No. 4 are at the Ranch House. You will remember the coulee by Rodgers house where we looked at the windmill. the water there was 60 feet wide and seven feet deep, it was as bad or worse at the dams.

I must close up Dam No. 1 as soon as it is repaired. we lost very little dirt from that, but it is not in good shape at bottom end, then wash what little with a small head into it. If I use a large head we shall lose good dirt which has hung up between 3 & 4. I wish you were here to see and help me consider the best thing to do, but it is far more important under the circumstances that you make your trip and do your every utmost to get the lower class goods moving so that we can get returns from all that material on hand. The better grades are not nearly so important as the output cannot be large, but please do push lower grades. This means an outlay of $10 to $15 Thousand dollars to repair, and I should consider the loss in repairs loss of time, dirt being moved to where it cannot be washed without great added expense, etc. Mine, Slum ditches and all told $75,000 up. Do not cut your trip short, make it more thorough and longer rather than shorter, but plan to spend some time here on your return.

Tollgate Hill is a peach, had County Commissioner up, can do the work and collect. . . .

Please dont let my letter discourage you, but encourage you to push low grade, it can and must be done.

Just a few lines as to how things look to you in that line would help at this time.

Yours sincerely,

(C. T. Gadsden)

Water from Coulee broke flume all watter from ditch and cloud burst combined down Tollgate hill some road

Gadsden then wrote a second letter, this one to the Syndicate directors in London, informing them of the situation at the mine. When it was finished, he gave it to Maude to enclose with a sapphire shipment she was preparing to go out that day. It would be one of the last large sapphire shipments ever to be sent from the English Mine.

July 28, 1923

My Dear Mr. Finigan,

I have received your letter of July 5th, and am enclosing corrected consignment (illegible).

I shall not be sending any consignments for some time. we have a 2-½ hour rain which has knocked me out of any decent output for this season. I think it will take (illegible) days at least before I can begin washing again and cannot wash out the balance of Dam No. 1 as Dam No. 2 must be rebuilt and refloored first. It took 3 dams, 850 feet of flume with iron bottoms completely demolished, most of the logs jammed up against the Ranch house. We cannot repair all the damage

The aftermath of the cloudburst and flooding that occurred on July 26, 1923. This photograph, taken shortly after the disaster, shows the severe damage done to holding dams and weathering floors that effectively terminated production at the English Mine.

this year. If we wash at all from now on, you can expect only a small output from now, and much calls for money, as usual. . . .
Must close now with kind regards in which Mrs. G. joins.

Yours sincerely,

(C. T. Gadsden)

But there would be no money coming to rebuild the mine and Gadsden knew it. Nor would any prospective purchasers of the property seriously consider the $150,000 asking price. After the cloud burst, Gadsden laid off half of his force and spent the remainder of the season washing what little dirt he could in makeshift sluices. The value of the 1923 production fell to $40,000, only one-fifth that of 1921. Without funds, Gadsden could only rewash "old" dirt. Mining ceased and the value of the 1924 production dropped to $25,000. The last year of recorded production was 1927, when Gadsden and a half dozen remaining workers washed out 6,451 carats of gem sapphires and 83,000 carats of industrial material. The entire 1927 production, the last ever recorded at the English Mine, was valued at a mere $4,850. For the first time in forty-nine years, the old Yogo mining district was left without a single producing mine. With a worldwide depression looming on the horizon, the markets for gems and jewelry were already drying up. The directors of the New Mine Sapphire Syndicate, with no hope of sale or production, officially declared the English Mine closed in 1929.

The discovery of the Yogo sapphires and the development of the Yogo dike into one of the world's major precious gemstone mines would be one of the more important, yet least remembered, aspects of Montana's frontier history. Although the turn of the century brought the end of the frontier to much of the West, the essence and spirit of the frontier seemed to survive longer in Montana. Most of the pioneers who played a role in the booms and busts of old Yogo had already passed on, but others were just now in their final years. Among them were two of the most colorful—Jake Hoover and Charlie Russell.

Jake Hoover, recognized as the discoverer of the Yogo sapphires, had moved on to Alaska in the early 1900s. The Klondike strike was already history when Jake Hoover and his partners, known as the "Montana party," arrived on Alaska's Kenai Peninsula, and headed for the Sunrise Mining District. Legend, which always seemed to follow Jake, tells that he struck a bonanza worth $50,000, but evidence is lacking. In 1907, he did stake a number of claims along a creek flowing into Cook Inlet, but apparently his golden dream eluded him again. In 1908, Jake had packed up and was heading south.

This time, Jake settled in Seattle where his half-sister lived. Although already sixty years old, he embarked on a new career as a professional salt water sport fishing guide. Jake lived and worked out of a boat house on Seattle's Lake Union, quickly establishing a reputation as an expert guide. "I furnish everything from boat to bait," read Hoover's 1915 advertisement in the Seattle city directory, "best fishing grounds in Seattle."

Montana may have heard little of Jake Hoover, but his old Judith

River Basin partner, "Kid" Russell, had become a genuine Montana folk hero who many considered the premier artist of the frontier. Since the early 1890s, Charlie Russell had made his home in Great Falls.

In 1922, Jake Hoover, apparently still plagued by his predilection for women and liquor, had run into legal difficulties over a serious offense involving a young woman. At age seventy-three, too old to continue his fishing guide business, Jake could not afford legal assistance. Proud as he was, there was only one person he could turn to for help—his old partner, Charlie Russell. Hoover's attorney sent a telegram to Russell in Great Falls, asking in Hoover's behalf for financial assistance. Although Russell could have known nothing of the details of the charges against Jake, he defended his partner to the end, recalling Jake Hoover as he remembered him four decades earlier in the hunting camps of the Little Belt Mountains.

> I am verry sorry to here that my old friend is in trouble and am more sorry that I cant do aney thing for him. I have no money and a broke friend is mighty little comfort to a man in Jakes fix. I have known Jake 42 years and in days gon I camped with him many years every man knows that camp fires and wether make men acquainted and I camped with Jake when it was 50 below and the sky was our roof Iv seen him tryed out and I dont beleve he wronged aney girl the big hills dident make that kind of men and Jake was a mountain mans man

Jake Hoover had never given up his dream of gold that first brought him to Montana Territory as a sixteen-year-old boy. Newspaper reporters often appeared at Jake's boat house seeking stories of "the old days." Jake's tales may have begun with hunting, but they always ended with gold. In 1923, he told an interviewer:

> You can tell them in Montana that I'm coming back there next fall, and that I'm going to make a few more strikes before I cash in. I know where there are some gold leads that will make people's eyes bulge out. That state has not been half prospected yet—not even for gold, which is the easiest of all the rich metals to find.

In October, 1923, a time when Charles Gadsden was still sorting through the ruins of the English Mine on the Yogo dike, Jake Hoover made his promised return to Montana to visit Charlie Russell in Great Falls. His vow to make that one last gold strike, of course, was only a brave dream, for Jake Hoover's prospecting days were far behind him. It is not known if Jake visited Utica, Yogo Creek or old Yogo City, the places near his old hunting grounds. If he did, he would have seen that his beloved cabins in Pig-Eye Basin were part of the dude ranch mentioned in the English Mine prospectus of 1922. This was Jake's last trip to Montana. Soon after returning to Seattle, his health failed. On April 25, 1924, news of the old hunter's death appeared in the *Great Falls Tribune*.

SAPPHIRE MINE FINDER DIES ON WEST COAST

DEATH OF JAKE HOOVER RECALLS EARLY DAYS IN TRIBUTE BY CHARLIE RUSSELL

"He was good to me when I was a boy," is the tribute paid to the memory of Jake Hoover, one of Montana's earliest pioneers and discoverer of the Yogo sapphire mines near Utica, Friday night by Charles M. Russell, Montana's cowboy artist, when he was informed of the former's death. Mr. Hoover died at Seattle, Wash., on April 14, death resulting from a stroke of paralysis which he suffered three months ago. Mr. Russell is one of a few pioneers living here who was well acquainted with Mr. Hoover, having made his home with him from 1880 to 1883.

At that time, Mr. Hoover was hunting and trapping, supplying meat to the settlers along the south fork of the Judith River and selling the furs and hides in Fort Benton. It is believed that he first came to this state in the early 60s from Skunk River, Iowa, being attracted here by the mining activities at Bannock and Virginia City. He followed prospecting in Fergus County, discovering placer gold at Yogo in 1879. "Buck" Buchanan was partner with him in this discovery. In about 1900 Mr. Hoover discovered the

> present Yogo sapphire mines at the old toll gate which guarded the first road into Yogo. A few years after the discovery of these mines he sold his interest to the late Matt Dunn of Great Falls, George Willis (Wells), and the late Senator S. S. Hobson, also of Great Falls. In 1914, these mines together with other mines owned by H. O. Chowen of Great Falls were purchased by the New Mines Sapphire Syndicate of London, England, in whose ownership they are in at the present time.

Years earlier, Charlie Russell, saddened by the inevitable passing of the Montana frontier, penned a simple poem for the men who had tamed it, from gold miners and gamblers to bull whackers and stage coach drivers. The "man with the gold pan" was undoubtedly Jake Hoover.

Here's to the first ones here, Bob,
Men who broke the trail,
For the tenderfoot and booster
Who came to the country by rail.

Here's to the man with the gold pan,
Whose heart wasn't hard to find,
It was as big as the country he lived in,
And as good as the metal he mined. . . .

So here's to my old time friends, Bob,
I drink to them one and all,
I've known the roughest of them, Bob,
But none that I knew were small.

Here's to hell with the booster,
The land is no longer free,
The worst old timer I ever knew,
Looks dam good to me.

Another of those old timers passed away in 1925—Jim Ettien, the sheepherder who took the time to pan the gopher diggings on the bench lands above Yogo Creek and thus discovered the Yogo dike. The next to go, on October 24, 1926, was Charles Marion Russell, the man who had done so much to preserve forever the soul of the

Montana frontier in pen, charcoal, oils and watercolors.

It is unlikely that Jake Hoover ever knew, or perhaps ever cared, what the ultimate value was of those blue pebbles he sluiced out of Yogo Creek in 1894 and 1895. For the record, Jake's discovery had been one of the biggest bonanzas in the West. By the time Jake died in 1924, the English Mine production on the Yogo dike had amounted to well over 16 million carats—*three and one-half tons*—of sapphires valued in the rough at *$2.5 million*. About fifteen percent—2.5 million carats—were of gem quality. Assuming a conservative cutting retention of thirty percent, the English mine had therefore provided the world gem markets with some 675,000 carats of fine, cut blue sapphires. With an average retail value of $30 to $40 per carat, the gem value of the Yogo sapphires had certainly exceeded *$25 million*. Jake Hoover, who confessed he "always sold out at the wrong time," had sold a quarter share of that fortune for $5,000. Jake, who would tramp half the continent searching for his bonanza, had his hands on it all along at the Yogo dike, but never knew it.

After the old timers were gone, the English Mine, by virtue of its official closure in 1929, itself passed into history. Charles T. Gadsden, who had already spent thirty long years at Yogo, by his own choice, remained with his wife, Maude, at the lonely mine. Once the resident supervisor of the greatest precious gemstone mine in North America, Gadsden was now the resident caretaker of his private empire, now only an empire of memories.

The official shutdown of the English Mine received little attention in the newspapers, due to the fact that the closure had really occurred over a period of several years and that newspapers, in the old frontier tradition, were still reluctant to print much bad news about mining. If the mine shutdown was mentioned at all, it was referred to as a "temporary suspension" of production.

On January 7, 1930, the *Great Falls Tribune* printed a philosophical editorial about Yogo. It was based on Jim Ettien's discovery of the Yogo dike in 1896, but would have been the ultimate epitaph for the grave of Jake Hoover.

GOPHERS AND BLUE PEBBLES

The story goes that on a dismal day in 1896, Jim Ettien, a sheepherder, was attracted by blueish pebbles at a gopher hole, 10 miles east of Yogo Creek in what is now Judith Basin County. And so it was that the despised gopher revealed one of Montana's greatest secrets. The gopher in burrowing his home had hinted at a treasure that has carried Montana's fame across the sea, for Yogo sapphires are now among the finest gems in the world.

. . . Ever since 1879 prospectors had sought gold in the lower reaches of the Little Belt Mountains, but substantial success was lacking. Of all the men who were in that area, it is probable that least of all Jim Ettien, the sheepherder, thought of an Eldorado. And it was he that chanced upon a secret that none of the other prospectors suspected. What his discovery may have meant to him, we do not know. But it is likely that Jim Ettien remained a sheepherder.

And so it is. Often the treasures of life are near at hand—so near that we do not suspect their presence. And when we do seek the treasures, usually we set out for far places, disappointment and failure being the recompense. Especially this is true of happiness, peace and contentment. We may search the world over for them to realize too late that we left them behind. As often as not we find them in the routine of our lives—we stumble upon them as Jim Ettien stumbled upon those blueish pebbles. And figuratively, it is the humdrum—the gopher—of life that digs up the treasures for us. . . .

Perhaps all of us who are old enough to have drunk of that deeper wisdom that comes only from the years, perhaps all of us who can meditate on the fundamental significance of things, will realize that if we weigh it through that the real treasures that have come to us in life are the blue pebbles that the gopher dug from the earth—not the treasures we sought in far places.

Looking west across Yogo Gulch into Kelly Coulee. The muddy area at the base of the Gulch are the tailings ponds of the Intergem operation at the American Mine site.

A tour bus parks before a mined-out area of the Yogo dike. This section was mined by the British in the early 1900s. This area shows "classic" dike formation, that of "pure" dike rock between two smooth walls of Madison limestone.

The Gadsden House as it appears today. The dike is visible on the hill to the left.

Remains of old workings of the English Mine.

An Intergem mine foreman inspects the concentrate recovered in the jigs.
Intergem, Inc.

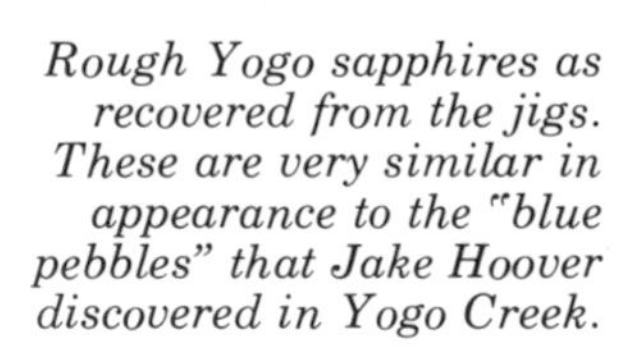

Rough Yogo sapphires as recovered from the jigs. These are very similar in appearance to the "blue pebbles" that Jake Hoover discovered in Yogo Creek.

Separating rough sapphires from the jig concentrates. Intergem, Inc.

The Tiffany Iris Brooch which contains 120 Yogo sapphires. Manufactured by Tiffany & Co. in the early 1900s, the brooch is the most elaborate piece of Yogo sapphire jewelry ever made. Intergem, Inc.

A $25,000 cocktail ring featuring a 2.80-carat oval-cut Kashmir blue Royal American Sapphire (Yogo sapphire) set in a ballerina style mounting surrounded by diamonds. Intergem, Inc.

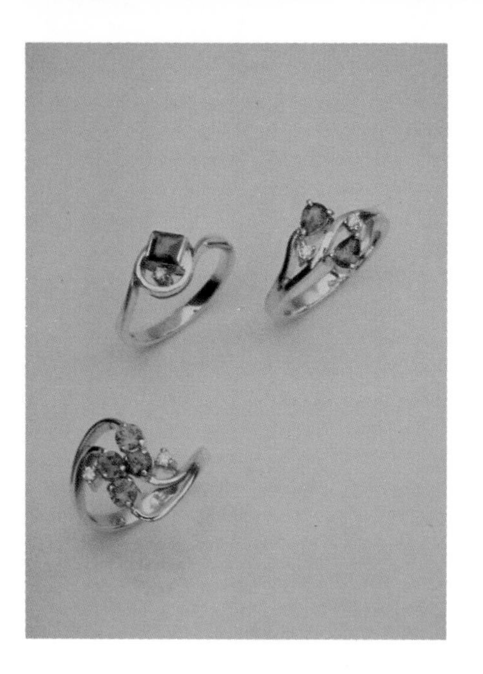

Yogo sapphires are the Royal American Sapphires appearing in Intergem's line of fine sapphire jewelry.
Intergem, Inc.

The Yogo sapphire is one of the most beautiful colored gems in existence. The violet stone is one of the two percent of all Yogo sapphires that are not blue.
Intergem, Inc.

The Yogo sapphire had been discovered by Americans, but exploited by the British who realized a fortune by mining only an infinitesimally small part of the blue treasure that lay locked in the Yogo dike. Now, as the British tenure was nearing an end, Americans, again, would have their chance. But first they would have to, in their own way and at their own pace, rediscover and reconfirm the worth and beauty of the Yogo sapphire. Tha search would prove longer and more difficult than merely panning blue pebbles from the gravels of a creek. Incredibly, it would take all of the next half century.

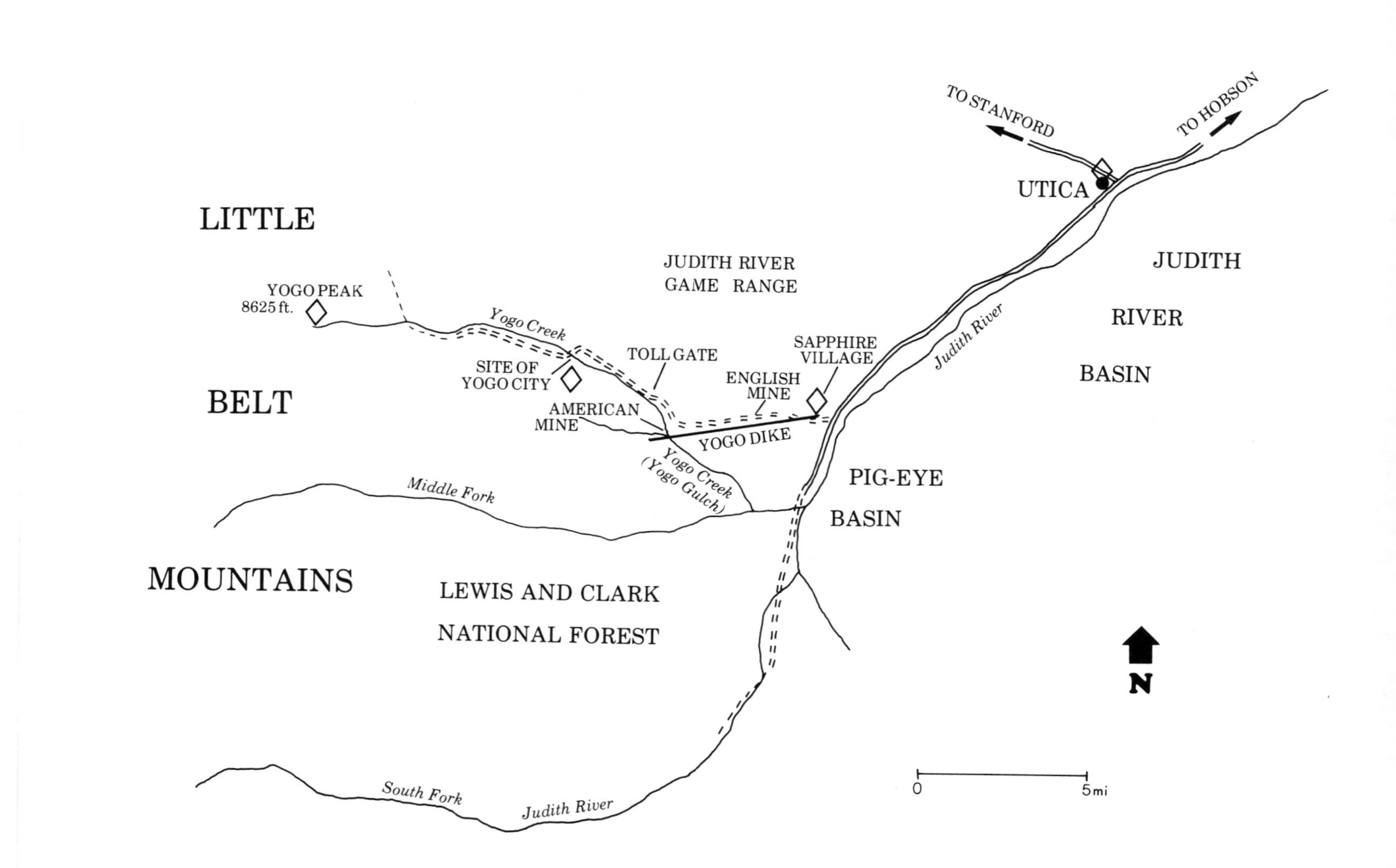
TO STANFORD
TO HOBSON
UTICA
LITTLE
BELT
MOUNTAINS
JUDITH
RIVER
BASIN
JUDITH RIVER
GAME RANGE
YOGO PEAK
8625 ft.
Yogo Creek
SITE OF
YOGO CITY
TOLL GATE
SAPPHIRE
VILLAGE
ENGLISH
MINE
AMERICAN
MINE
YOGO DIKE
Judith River
Yogo Creek
(Yogo Gulch)
PIG-EYE
BASIN
Middle Fork
LEWIS AND CLARK
NATIONAL FOREST
South Fork
Judith River
N
0
5mi

Chapter 4

All Is Not Gold That Glitters

Fears that synthetic gems would undermine the natural precious gemstone markets had proven unfounded. Natural stones, because of their inherent rarity, would always hold or increase their value, while synthetics, worth much less, would fill a market for beautiful, high-quality gem substitutes. Precious gem markets, although not vulnerable to the flood of synthetics, staggered under the economic onslaught of the Great Depression. All over the world, what little money was available went for the necessities of life, not for the non-utilitarian luxury of gems and jewelry. Not even the finest blue sapphires had a market; ironically, however, inexpensive clear sapphires had come into considerable demand. By the thousands, diamonds were removed from their mounts in family jewelry, to be replaced by clear sapphires known cynically as "Depression diamonds."

Events at Yogo reflected the depressed world gem markets. Charles Gadsden now ruled an empty kingdom; with no production and no employees, he and Maude lived increasingly reclusive lives at the remote, lonely mine. Contact with the Syndicate directors in London was much less frequent, but Gadsden did not seem concerned. With their gardens and livestock, he and Maude had become quite self-sufficient. Gadsden visited Utica, Lewistown or Great Falls only to buy provisions or to transact what little business was necessary. At one time, Maude remained on the mine property for three straight years. Even though Gadsden had been a naturalized American citizen for thirty years, he never allowed himself to become Americanized, even to the point of listening only to Canadian radio stations. Domer L. Howard, a writer who met Charles Gadsden during the 1930s, described the Englishman as he appeared during those years.

> Gadsden could hear a car coming up the rough, winding road while it was still a half mile away and he would be sitting on an ore heap with a .30-30 rifle across his knees when the occasional visitor arrived. With a courteous dignity bordering almost on shyness, nevertheless one look at the firm set of Gadsden's stubborn British jaw with its overhanging black mustache, and a few words of his sharp, incisive Cockney accent, quickly convinced the visitor that this was no man to trifle with. Although the Gadsdens accorded a civil courtesy to visitors, they always maintained enough English aloofness and formality to inhibit any close friendships.

Even though the mine was not producing, Gadsden kept himself quite busy, tending gardens and livestock, maintaining equipment and regularly checking the entire five-mile length of the Yogo dike. Several years earlier, Gadsden had branched out into mink raising and, by the early 1930s, become a mink breeder of regional note. The newspapers, always interested in news from Yogo, found Gadsden's diverse interests a suitable substitute for sapphires, as shown in this article from the December 21, 1931, *Lewistown Democrat News*.

YOGO GULCH MAN IS DIVERSIFYING

MINER FINDS TIME TO PRODUCE SAPPHIRES, RAISE MILK, GOATS AND MINK

Sapphires, mink and milk combine to keep one Montana mine operator busy through the year. An account of this unusual diversification of industry was brought back to the Montana School of Mines by Prof. Eugene S. Perry, following one of his investigation trips for the state bureau of mines and geology.

Supt. C. T. Gadsden of the Yogo gulch

> sapphire workings at Utica is the man who divides his attentions among three such unique occupations. . . .
>
> When sapphire production is curtailed because of market conditions, Gadsden devotes his time to raising fancy breeding mink, as well as to giving expert attention to odd mining jobs about the property. Like Yogo Gulch sapphires, his mink, dark-colored and free from spots, are famous and have won blue ribbon awards at Seattle and Montreal shows of fur-bearing animals.
>
> Like Mahatma Gandhi, Gadsden thinks goat's milk is delicious to drink and keeps a few prize goats. Formerly, he took much interest in raising them, also, but his attention has recently been turned more to mink.

Gadsden's mink represented more than interest and enjoyment, but a vital source of income. Since the events of the 1920s had ended production at the English Mine, the arrival of Gadsden's wages had become sporadic. The Syndicate directors now found that, instead of owning a money-maker, they were saddled with a liability. The mine was non-producing, the property was severely flood damaged, no prospective buyers were coming forth, the value of the Syndicate shares was questionable, yet certain minimal expenses had to be met. By 1933, the arrival of the funds to cover even maintenance expenses and property taxes became irregular. The directors were turning their backs on the mine, letting their faithful resident supervisor fend for himself.

It would seem that Gadsden had the option to walk away from Yogo, perhaps to return to England. But, in 1934, Gadsden was already sixty years old and had spent thirty-five years—over half his life—at the English Mine. While he had proven himself as a supervisor, his experience had prepared him only for managing another gemstone mine, a profession with little demand in the 1930s. Furthermore, Gadsden's personal assets were based almost solely upon his holdings of New Mine Sapphire Syndicate stock. For that stock to be worth anything, the mine would either have to produce again or be sold for a substantial price. Meanwhile, there were fuel bills, telephone bills, and living and maintenance expenses to be met. The biggest concern was property taxes. If the tax payments to the county stopped, the consequences would be unthinkable—the property might well be seized and sold for a pittance at a Sheriff's sale. Much of the proceeds of Gadsden's mink business went into payment of property taxes. He still washed some sapphires from the heaps of deteriorated dike rock that lay about the property, occasionally selling some of the better stones in Lewistown and Great Falls to supplement his income. Where seventy men once mined and sluiced, one aging Englishman was left to slowly shovel deteriorated dike rock. The success of the English Mine over a quarter-century had been the crowning achievement in Gadsden's life, but it had also become an obsession, the very reason for his existence. Gadsden had become a slave, of sorts, to the ruined mine. He could not and would not leave Yogo.

With the English Mine inactive, the story of the discovery and development of the Yogo dike was becoming steadily displaced by legend and lore. With most of the old timers gone, writers recreated the history with considerable license. Although most accounts were at least loosely tied to the facts, some, such as this discovery version which appeared in a 1938 magazine, were closer to fiction.

> The Yogo Lode it was called, as Yogo was the Indian word for sapphire, and it was a mine that the Indians had fiercely guarded for centuries. It was up to a miner named Jack (sic) Hoover to make the great strike in 1876. Hoover, an Englishman who went west with the Fortyniners, had an unerring eye for anything precious and first struck a gold lead. With pick and shovel, Hoover followed the rich lead, filling his pan as he went, until he came to even a richer treasure—sapphires! The greatest heap of glittering blue sapphires man had ever seen!

The Montana newspapers, which printed lengthy historical "inserts" about the past glories of Yogo, were also guilty of inaccuracies. Rarely is the Yogo story told through the same set of names, dates and events. Still, the Montana newspaper writers could be quite protective of their Yogo history. In 1940, a *Great Falls Tribune* writer took an eastern newspaper to task over the details of the discovery of the Yogo dike.

> Some time ago I read an interesting account of the discovery of the Yogo sapphires in The Forward, a paper published in Philadelphia

which attributed their discovery to the burrowing of ground squirrels in a cornfield. As corn is not a staple crop at the elevation of 5,000 feet in the Belt Mountains, we must accept this account as an example of how accurate some eastern writers are on the subject of Montana history.

By 1939, the only sapphires produced at Yogo in twelve long years were those washed out by Charles Gadsden for his own personal disposal. Still, the earlier profits and potential of the Yogo dike were not completely forgotten. As world gem markets began a slow recovery after the Depression, interest in reactivating the English Mine appeared. Interestingly, it came not from the United States, but from London where another company was considering taking up where the New Mine Sapphire Syndicate left off. Montana read of the news in the *Helena Herald* on February 13, 1939.

ENGLISH FIRM WOULD LIKE TO DIG FOR MONTANA SAPPHIRES

Foreign capital has been making inquiries about the possibility of getting into the sapphire industry in Montana. Dr. Francis A. Thomson, president of the Montana School of Mines in Butte, made an announcement to that effect at the recent All-Montana Industrial banquet at Helena.

The firm considering going into business is the New Mines, Inc., of London, which wrote Dr. Thomson some time ago asking if Montana was interested in "foreign" capital or if it were barred in the United States. The London firm was somewhat interested in developing the properties in Yogo Gulch. The School of Mines President wrote back to London and said that new capital was always welcome.

Sapphire mining in Montana has, in the past, been a fair-sized industry and in the early days English firms took all the Montana sapphires they could back to London, cut them, polished them, and sold them as Indian sapphires, because those were generally considered to be tops. But by and by one person and another got to noticing that the Montana stones were quite a little better than those produced in India and eventually the Montana sapphire was able to stand on its own merit in the markets of the world. In recent years, however, the industry has declined but, if the letter from London is any indication, maybe it's due for a comeback.

The seriousness of the renewed British interest in Yogo was never determined, for the outbreak of another world war diverted any and all interest, effort and capital. Although World War II put another damper on world gem markets, it created a skyrocketing demand for bearing jewels and industrial corundum. Synthetic corundum was now used by most bearing jewel manufacturers, but natural corundum and industrial grade sapphire was preferred, because of cost, for abrasive and cutting purposes. On April 13, 1941, the *Great Falls Tribune* reported the sudden new importance of the Montana sapphire deposits.

WAR MAY BRING NEW DEMAND FOR STATE SAPPHIRES

HELENA, April 12—(AP)—The Atlantic blockades, seeking an even tighter stranglehold on transoceanic commerce, may steer the western world to Montana's doorstep for sapphires.

Uncle Sam himself may bring the biggest customers.

Suggest "sapphire" to a girl (or boy) friend and the jeweler's might be the next stop. But sapphires have a different, more important value in national defense.

Natural sapphires are excelled only by diamonds for hardness of composition and resistance to wear. Because they are much cheaper than diamonds, natural sapphires are widely used for bearings in scientific apparatus of many kinds, including gauges, scales, electric power me-

> ters, compasses, and airplane and navigation instruments. . . .
>
> Then came the war with its British blockade and the German counterblockade. American industrial firms—particularly large electrical manufacturing concerns—contributed to a new domestic market which expanded until in 1938 Montana's small placer claims were producing sapphire for sideline income.
>
> Sensing the trend, Montanans, Inc., the state chamber of commerce, prodded the office of production management in Washington for an opinion on the application of Montana sapphires to the national defense program.
>
> One result was a letter from Alexander Shayne, a specialist on jewel bearings for the metals and minerals section of the OPM. The letter noted that natural sapphires are most widely used for industrial purposes, while synthetic stones go to watchmaking and said:
>
> "Our survey indicates that, for the first category of jewels, natural sapphire is preferred by the industry and various technical papers on the subject indicate the preference lies with the Montana sapphire when properly selected. . . .

The wartime demand for corundum and sapphires instilled temporary life into two historic mining districts—Rock Creek and the Missouri River bars. By 1944, when activity ended, these two deposits had produced over seven million carats—two tons—of sapphire. Although tens of thousands of gem quality, vari-colored sapphires were recovered, the entire production was manufactured into jewel bearings or abrasive and cutting materials. The importance of Montana's wartime sapphire production was reflected in one of the very rare exceptions ever to Executive Order L-208, the presidential decree that closed all gold mines so that men and equipment could be directed to production of strategic metals. The Perry-Schroeder Mining Company, of Helena, was permitted to continue operating its gold dredge on Eldorado Bar. Even though the primary economic product was gold, the co-product recovery of sapphires was judged vital to the war effort. If the Yogo dike had been in American hands, it could easily have doubled Montana's wartime sapphire production.

Jewel bearing and industrial uses for natural corundum were in their last years. Synthetic corundum had already replaced natural sapphire and ruby in most bearing jewel applications. Meanwhile, research was well along in the synthesizing of silicon carbide, an abrasive material with a Mohs hardness of 9.4, much harder than natural sapphire and approaching that of diamond. In the future, the non-gem uses and value of industrial sapphire would be very limited. If sapphires would be mined at all, they would be mined for their gem value only.

After the war, the United States began a post-war industrial boom that would result in record levels of prosperity. Among the many markets to make strong recoveries would be gems and jewelry. Accordingly, interest again focused on the Yogo dike where Charles and Maude Gadsden still faithfully watched over the property. For the first time, the British were not interested in Montana sapphires. World War II had devastated the pound sterling. Not only were the British desperately trying to stem their foreign expenditures, they had begun cashing in many of their foreign assets to acquire hard currencies. Now, the interest in the Yogo dike was American. A growing number of automobiles, each trailing a cloud of dust, could be seen driving up the old road from Utica. The visitors all had similar purposes: meet Charles Gadsden, tour the Yogo dike and, with an eye to the future, survey what was left of the old English Mine workings.

Among the visitors was Stephen E. Clabaugh, a United States Geological Survey field survey geologist, who was conducting a general survey of Montana's corundum deposits. Since the Yogo dike was the state's premier deposit, Clabaugh and his wife spent several weeks with Charles and Maude Gadsden during the summer of 1946. Clabaugh's report, *Corundum Deposits of Montana*, was available in United States Geological Survey offices in 1947 and was finally formally published as United States Geological Survey Bulletin 983 in 1952. The report was valuable in that it generally documented the discovery, production and development of the English Mine. Ironically, its greatest value would not be to geologists or mining men, but to promoters who would quote from it for decades to come.

> . . . The Yogo sapphire deposit is the most important gem locality in the United States. Cut sapphires, of excellent quality, valued possibly at as much as $20,000,000 to $30,000,000 have been produced from the deposit, and reserves of sapphire-bearing material are probably adequate to supply several times the quantity mined. . . .
>
> The geology of the dike is exceedingly simple. . . .
>
> A minimum estimate of reserves of altered dike rock is therefore considered to be about twice the amount removed, or approximately one million tons of material containing more than 25 million carats of sapphires, of which some 4 million are gem quality. The dike undoubtedly extends downward and to either side several thousand feet beyond the area considered in this estimate, hence a figure several times greater than the estimate given above is not unreasonable.
>
> Small colored gems of all types have been much in demand in recent years in the manufacture of fancy jewelry in the United States, and the value of the natural sapphires has continued to rise, in spite of the production of excellent synthetic stones. Therefore it appears reasonable that an American company might acquire and profitably work the Yogo deposit, producing gems for domestic consumption. . . .

Nowhere in Clabaugh's report was any inference that a king's ransom in sapphires was waiting to be shoveled out of the Yogo dike. The report was a general geological survey, a statement of the value of previous production, an estimate of reserves, mention of the strengthening gem market and prices, and a personal assessment of the future mining potential. Taken together, Clabaugh's words were a promoter's dream. Bulletin 983, written under the aegis of the United States Geological Survey, had not only instant and unquestioned credibility, but an inference that went far beyond its intended meaning. Promoters and future sapphire miners had found a "bible" in Clabaugh's report; with it, they had no need or use for geological or mining surveys of their own. No one would be able to guess how many times United States Geological Survey Bulletin 983 would cross a desk between a promoter and a prospective sapphire mine investor in the next three decades. It would be the hook that would raise hundreds of thousands of dollars. Skepticism and caution on the part of a potential investor could easily be overcome by opening United States Geological Survey Bulletin 983 to the appropriate page and saying quietly, "Yes, but remember, the United States Geological Survey says. . . ."

The first American company to turn to the Yogo dike was the Yogo Sapphire Mining Corp., a group of Billings, Montana, men led by

Thomas P. Sidwell. Their initial offer, made in 1946, matched the 1922 asking price for the New Mine Sapphire Syndicate—$150,000. Several years were required to track down a quorum of the surviving Syndicate shareholders in Britain. By 1949, there was agreement, at least in principle, to take up the American offer. Yogo, after an absence of thirty years, was suddenly current news again. On October 6, 1949, the *Lewistown News-Farmer* ran the first article about the "new" Yogo.

YOGO SAPPHIRE MINES WILL REOPEN NEXT SPRING; MAY BE BOON TO CENTRAL MONTANA

. . . the Yogo mine will go into operation this coming spring with the most modern equipment available, Thomas P. Sidwell of Billings, one of the directors of the Yogo Sapphire Mining company, announced Friday.

The new corporation has purchased all the property and equipment from the New Mine Sapphire Syndicate. The transaction was completed in September after nearly two years of negotiations.

The opening of the mine will begin a new era in the history of Central Montana. Sidwell, who is general manager of the mine, said that the stones, among the finest in the world, will be advertised internationally. He predicted that not only will the mine be a profitable venture, but will also bring national advertising to Lewistown and Central Montana, and visitors from throughout the country will visit the mine.

Mr. and Mrs. Charles T. Gadsden will remain at their home at the mine site. Arrangements have been made, Mr. Sidwell said, for them to stay as long as they live. The property has been supervised since 1889 (sic) by Mr. Gadsden and Mrs. Gadsden came there in 1903. They have

> returned to England several times but otherwise have stayed close to the mine, showing visitors through the property. Last summer, 366 visitors registered in Gadsden's guest book, despite the fact that the mine had been out of the limelight for a number of years.
>
> Approximately 30 to 40 men will be employed at the mine next summer. New equipment will be brought in and renovation of the mine will begin just as soon as the snow goes off the ground, Sidwell said. A new shaft will be dug and buildings will be constructed for the workmen.
>
> Frank Bryant of Lewistown, who has been interested in the mine for many years, will be the engineer in charge, and Lloyd Wartes of Spokane, who has made a thorough study of the South African diamond fields, and John Abrams of Los Angeles, will be the consulting engineers. . . .
>
> Under the new management, arrangements have been made with a large jewelry corporation on the west coast to have a world-wide market. . . .

The Yogo Sapphire Mining Corp. was the sixth American company to appear at the Yogo dike. John Burke and Pat Sweeney, original claim holders at the American Mine, and some of the original partners of the New Mine Sapphire Syndicate, had sold out at handsome profits. All of the others had found only frustration and bankruptcy. Each had also begun desperately trying to raise capital, and the Yogo Sapphire Mining Corp. would be no different. Although Sidwell's corporation had an authorized capital of $1 million, very little hard cash had actually been raised. That job, hopefully, was to be taken care of by a slick, promotional brochure titled *Yogo Sapphires–The Story of Montana's Marvelous Treasure*. After the usual presentation of discovery and English Mine development, the brochure took dead aim at investors' wallets, breathlessly listing the long-term profits accrued by the New Mine Sapphire Syndicate, then explaining how modern American mining technology and marketing methods were the sure key to more.

DIVIDENDS OF 317%

The new British owners were not expert mining men. As early as 1907, an American mining engineer wrote that their methods were "old fashioned"—and that the owners themselves were "stubborn and hide-bound." So, the working shaft which was sunk, the steam hoist which was installed, the other improvements that were made were far from elaborate. But the results were fantastic! (If the reader may allow us that bit of superlative expression.)

In the 22 years between 1901 and 1923, the syndicate mined only 200,000 tons of gem-bearing material—hardly 5% of the total estimated material in the dike. In the same period, the British group paid for all the development work at the mine, bought out a rival company which shared their ownership in the dike. . . .

AND THEY STILL PAID THEIR STOCKHOLDERS A TOTAL OF 317% IN DIVIDENDS DURING THE 22-YEAR PERIOD.

Thus each stockholder received an average annual yearly dividend of almost 15% on his investment—or a return of better than three times his total investment in 22 years!

A NEW DAY IS DAWNING!

It was in 1947 that a small group in Montana decided that the time was ripe to acquire control of the Yogo sapphire mine from its far away British owners.

The timing was superb!

This time the British were fighting for their very economic life—with no hope of expanding their foreign investments.

This time the British badly needed the American dollars which they would get from the sale of the mine.

And this time—the Montanans were determined that one of their state's richest resources should not continue idle.

The negotiations went forward—slowly, tediously, almost endlessly.

And finally in October of 1949, the purchase contract was signed!

Soon the gem markets of the world may again thrill to the sight of beautiful Yogo sapphires!

THE FUTURE IS BRIGHT!

Soon a dream can become a reality—in a plan which contemplates that:

Modern American machinery will start moving into place at the Yogo sapphire mine—and new dike material will be weathering to give a waiting world more of the incomparable Yogo sapphires!

But even without more mining it is estimated that thousands of dollars worth of sapphires can now be recovered from the dike material dug years ago. Mining engineers estimate that there are 4,800,000 TONS OF SAPPHIRE BEARING MATERIAL STILL UN-

> TOUCHED—and only 200,000 tons already mined.
>
> The new purchasers know that this Montana dike is believed to be the only one in the world carrying gem sapphires—and they know that sapphires have been found wherever this dike is opened.
>
> They know that modern American machinery to mine the sapphires, and modern American merchandising methods to exploit the world markets can bring results of which the former owners did not dream.

The "modern American machinery" was not yet about to rumble into Yogo, for the printed word was somewhat premature. Not only hadn't the actual transfer taken place, seven years of legal wrangling would precede it. In Britain, tracing the ownership of the 175,000 outstanding Syndicate shares, many of which had been issued a half century earlier, was a lengthy matter. At least 10,000 shares were considered lost, and many of the remainder had been passed on several times through heirs and estates. By 1950, the Americans and the British were at odds. The Yogo Sapphire Mining Corp. charged the Syndicate with not formally revealing its stockholders' decisions, then even questioned the "merchantability" of the property title. The Syndicate countered, charging the Americans with both breach of faith and contract in not depositing the required funds in the proper accounts.

Thomas Sidwell's personal contact in negotiations with the Syndicate, of course, was Charles Gadsden, who still lived in the house on the hill overlooking the Yogo dike. Fifty years had not tempered Charles Gadsden's dislike and distrust of Americans, and Sidwell was not about to become the first exception. Friction between the "stubborn" Englishman and the "pompous" American, as they referred to each other, did little to facilitate negotiations. It is not likely that Charles Gadsden would ever have considered selling the property to Thomas Sidwell, but that decision was not his to make. Since the Syndicate had no intention of ever working the property again, Gadsden knew that its sale to Americans was inevitable. He also knew that he had put $29,000 of his own money into property maintenance and taxes, and that the Syndicate had expressed little interest in paying it back. To protect himself in the likely event of a sale, Charles Gadsden filed suit against the New Mine Sapphire Syndicate in a Lewistown court. That lien further complicated the fragile negotiations in progress.

In 1952, Charles Gadsden was seventy-eight years old and in fail-

One of the last photographs ever taken of Charles and Maude Gadsden on the Yogo property. They are seated on the porch of the house that was built for them in 1913. When the house was built, it was one of the most modern in the region, boasting a steam boiler in the cellar for heat. The house also contained a sizeable vault for sapphire storage. The Gadsden House still stands today in a remarkable state of preservation.

ing health. Because of age, health, and a continued dislike for the would-be American sapphire miners, he and Maude finally left Yogo, their home of almost fifty years, and moved into a small house in Lewistown where, the following year, they celebrated their golden wedding anniversary. On March 11, 1954, Gadsden passed away. In the obituaries, Montana newspapers paid tribute to him as one of the state's pioneers. His remains were cremated in Great Falls, then flown to England to be interred in the family plot. In the fifty-five years since he had first stepped down from the stagecoach at Utica, the Englishman's name had become locally synonymous with the Yogo sapphire. Now, at last, Charles T. Gadsden had gone home.

Maude, who had been in poor health for several years, remained in Lewistown.

Gadsden seemed to have been the last British tie to Yogo. His passing was a catalyst of sorts to end the lengthy negotiations. In July, 1956, ownership of the Yogo property was formally transferred to the American-owned Yogo Sapphire Mining Corporation. The final agreement was $65,000 cash and various stock considerations to acquire 145,000 of the 150,000 traceable British shares. Some locals said that Charles Gadsden had gotten his last wish—never seeing the Yogo mine sold to Americans during his lifetime.

It had taken the Yogo Sapphire Mining Corporation a decade to acquire the Yogo sapphire deposit; now that they had it, their operating capital was exhausted. The corporation reorganized under the old British name—the New Mine Sapphire Syndicate. Instead of "modern American machinery," the American Syndicate could only move in a small crew to scrap old buildings and perform general cleanup work in preparation for small-scale mining the following spring. For years, the only Yogo news had come out of courts and attorneys' offices; the favorable publicity that appeared in the *Billings Gazette* on July 14, 1957, was a refreshing and encouraging change.

YOGO ORE PILES YIELD SAPPHIRES "CORNFLOWER BLUE" STONES ARE FOUND

Recent news from the Little Belt Mountain area northwest of Harlowton has shown the famed Yogo Mine has lost little of its potential for producing some of the world's largest and most beautiful sapphires.

Finds of 4½ and 5½-carat stones of the glistening "cornflower blue" hue particular to the area have been reported in the past week.

Discovery of the two large sapphires bolstered hopes of the owners and operators that the recently-revived mining venture may again produce a gem as

> large as the 19-karat (sic) gem found there.
>
> Don Johnson, a Billings policeman said he and Thomas Sidwell, a Billings and Lewistown resident who heads the company, were together at a sorting table when the 5½-karat (sic) gem was found. "The sun hit it and it just seemed to jump out of the gravel and dirt," Johnson said. . . .
>
> Still standing, battered and weather-beaten, are several buildings. Machinery is rusted beyond repair, mine shaft supports have tumbled and sluicing works have needed much repair.
>
> Johnson said a crew of ten men has been working in the area, utilizing old sluice boxes and runs and processing ore piles that have been standing dormant since 1927.

Few sapphires were sluiced out, for most of the work was directed toward tearing down old structures to improve the appearance of the property. The cleanup uncovered heaps of old records and correspondence that Charles Gadsden had painstakingly filed and which represented the complete history of the English Mine. Some were saved, but the bulk was unceremoniously tossed down one of the crumbling old shafts and bulldozed over. Along with the destruction of the old records, another of the last ties to the English Mine also passed into history. Maude Margaret Mellor Gadsden, after spending her last two years in a Lewistown rest home, died on June 21, 1958, and was laid to rest in a Lewistown cemetery.

The American Syndicate, strapped for capital and unable to mount a serious mining effort, contracted Siskon, Inc., a Nevada construction company, to work the mine in return for a share in the profits. Some new equipment was moved in, including a Judson-Pacific revolving trommel to replace the frontier-era sluice boxes. The trommel could wash up to 150 tons of deteriorated dike rock per shift, rejecting any oversize material while allowing undersize material to be passed to a series of gravity and sizing jigs for recovery of sapphires. Yogo made the front page of the *Great Falls Tribune* on September 1, 1959, in an article that smacked of stock promotion.

YOGO SAPPHIRE MINE PRODUCES $6,000 WORTH OF GEMS DAILY

LEWISTOWN—Central Montana's Yogo Sapphire Mine near Utica, the most important gem locality in the United States, is producing about $6,000 worth of gems a day since it began operations in early May, according to Thomas P. Sidwell, a director of the New Mine Sapphire Syndicate. . . .

The mine, which was established in 1898, had been controlled by a British syndicate until 1956, when Sidwell, representing the American Syndicate, bought 140,000 of the 150,000 English shares. The mine had not been working since 1927 when a cloudburst washed away its workings (sic).

A crew of five is working the mine, he said, and although it is impossible for so few to operate the mine at the most productive capacity, it has been producing about 3,000 carats of sapphires per day.

Sidwell said the syndicate is planning a new location for the Judson-Pacific washing plant which is at the bottom of Yogo Gulch on the banks of Yogo Creek. He said. "We aren't able to work to capacity because two dump trucks can't fill the machine fast enough." The trucks get their "payloads" two miles up the gulch from a five-mile-long dike and must haul it two miles to the washing plant in Yogo Gulch.

"With a larger and better located plant installed," he said, "we can produce up to 5,000 carats a 'day.' The new locations of the plant is planned further downstream at the site of the old American mine which went out of operation in 1914."

The printed figure of "$6,000" worth of sapphires per day was doubtlessly inflated, for, at that rate, a season's production would have amounted to a half million dollars. However much was produced, it was not enough to warrant any further mining after 1959. The actual sapphire production from 1957 to 1959 was probably much closer to 50,000 carats, of which 4,000 may have been gem quality. But even 4,000 carats of rough gem-quality sapphire, without a cutting and marketing system, weren't worth much at all, a lesson American sapphire miners had yet to learn.

The poor production was compounded by continuing legal and financial problems. The case of Charles Gadsden vs. The New Mine Sapphire Syndicate had not yet come to trial when Gadsden died in 1954. Maude, however, pursued the action, filing her own writ of property attachment against the British-owned syndicate. In May, 1958, a judgment was finally handed down for $29,000 against the American-owned syndicate. Although Maude Gadsden died only a month later, the executor of her estate carried on the litigation. Meanwhile, Siskon, Inc., dismayed over the absence of any financial return for their work during the 1959 mining season, filed a breach of contract suit against the beleaguered New Mine Sapphire Syndicate to recover $54,960 for work performed.

After mining was terminated in September, 1959, the mine property was left unguarded. For the next three years, the Yogo dike was the prime mineral collecting, or "rockhounding," site in North America. The first visitors were local ranchers, farmers and other county residents who came out of curiosity to see what remained of the fabled English Mine and to pick through the heaps of weathered dike rock. As word spread that the Yogo sapphire deposit was "open," they were followed by a small legion of rockhounds and mineral and gemstone collectors from all over the United States and Canada. They came with tents, campers, cars and pickup trucks and were equipped with everything from gold pans and sluices to sorting tables, screens, vibrating tables, "dry washers," and rock picks—even a few divining rods to help find "the big one." Some came for a day or two to roam over the historic site, others came prepared for a month-long or even a season-long stay to search through tons of dike rock for the legendary Yogo sapphires. During these years, Yogo was probably the only place in the world where anyone could walk about openly searching for precious gemstones. Some locals have suggested that, if Charles Gadsden was blessed by dying before the mine ever passed back into American hands, he was doubly blessed now.

No one will ever know how many Yogo sapphires were recovered by rockhounds during the three years the mine was "open." It certainly amounted to several thousand carats of cuttable gemstones, including individual specimens of three and four, even five carats each. Freelance sapphire mining reached such proportions that the New Mine Sapphire Syndicate was finally forced to take action in 1963. On April 15, this article, among similar ones in other Montana newspapers, appeared in the *Great Falls Tribune*.

**SAPPHIRE MINE
NOW CLOSED
TO SEEKERS OF GEMS**

LEWISTOWN—Rockhounds who journey to the sapphire diggings on Yogo Creek will be in for disappointment, according to the New Mine Sapphire Syndicate, which announced that all roads leading to the area will be blocked by locked fences.

The syndicate said that the area is closed to the general public and trespassers may be subject to prosecution. For the last few years, the abandoned mines have been a popular source of gem stones and thousands of amateur prospectors have searched for sapphires.

When Charles Gadsden still resided on the mine property, "No Trespassing" signs had never been necessary. Gadsden's reputation and the sight of the tall, somber-faced Englishman standing at the gate with a .30-30 cradled in his arm had been enough to deter the most determined rockhound. The "No Trespassing" signs that had been posted in the late 1950s had been taken down by collectors during the "open" years. The signs that now went up may have been a deterrent to some, but other rockhounds found them to be an intriguing indication that they were certainly at the right place if they wanted to find sapphires.

NO TRESPASSING

A REWARD OF $100.00 WILL BE PAID FOR INFORMATION CAUSING THE ARREST OR CONVICTION OF ANYONE STEALING OR DAMAGING ANY OF THE PROPERTY ON THESE PRE-

MISES. PERSONS, GEM COLLECTORS, AND/OR ROCKHOUNDS WILL BE PROSECUTED FOR POSSESSION OF AND/OR GATHERING AND/OR REMOVING SAPPHIRES OR ORE FROM THESE PREMISES.

The New Mine Sapphire Syndicate now began leasing the property to anyone who came along with little cash. At least two groups tried their sapphire mining luck, both briefly and unsuccessfully. But time had run out on the New Mine Sapphire Syndicate. After nineteen years spent in acquisition and futile development of the mine, the Montana Supreme Court, in May, 1965, ruled in favor of Siskon, Inc. On June 24, Siskon was permitted to purchase the Yogo property at a county Sheriff's sale held at the nearby town of Stanford. The property, consisting of seventeen patented lode claims and fifteen patented placer claims, now totalled 1,860 acres, or about three square miles. It was, as Stephen Clabaugh had noted in United States Geological Survey Bulletin 983, "the most important gem locality in the United States," and contained the entire known length of the Yogo dike with its uncounted millions of carats of gem quality sapphires. The price at the Sheriff's sale was $75,000—about $45 per acre. Even though the price took into consideration the court judgment on monies owed Siskon, it was clearly one of the greatest mining property bargains ever transacted in the West.

But Siskon, now the sole owner of the Yogo dike, wanted nothing to do with sapphire mining or the yet unfathomed mysteries of sapphire marketing. Again, the property was leased, this time to a group headed by Arnold Baron. Baron, with his background in jewelry and gems, approached Yogo with a realistic awareness of some of the problems that stood in the way of a successful commercial sapphire venture. Baron's expertise rested primarily in the fields of cutting and marketing. He recalled his experiences at Yogo in a talk delivered at the International Gemological Symposium in spring, 1982.

> A few years before this (1964) I had opened a small jewelry store in Montana. The mine was abandoned; every rancher in the area spent his time at the mine looking for sapphires. Some decent ones were recovered and a few of them found their way to our store for an opinion of their value. They were good enough to attract my attention. Two friends of mine in the oil business expressed an interest and the three of us agreed that if title to the property were ever resolved we should take

a closer look at it.

Shortly after the Supreme Court decision in 1965 we entered into a contract with the new owner, leasing the mine for two years with an option to buy it at the end of the lease. We were permitted to mine and market to evaluate the potential and defray our expenses. One of my associates, a geologist, took charge of the mining, one handled the finances and my responsibility was the final recovery from the concentrate and marketing.

First we considered the possibility of selling the rough. The main advantage would have been that the sale of rough is a cash sale. However, it didn't appear that we would be paid enough to allow for profitable mining. If one markets the cut sapphires, he must wait at least a year longer for his money, but since the anticipated return was considerably greater, we committed to the cutting program.

Baron evaluated the quality of cutting in several foreign countries, finally organizing a system of cutters in Germany and Thailand. For the first time in a half century, newly mined Yogo sapphires began appearing on the United States market. Baron recalled:

We were pleased with the results in cutting and marketing. At first we encountered more resistance to the Yogo in this country than we did abroad. . . . Some Americans thought it wasn't a sapphire unless it came from the Orient. Eventually the American consumer reacted very well to the Yogo and so did the dealers. I believe that during their years, the British must have marketed the best of the Yogos in this country as Orientals in the belief that these would bring a higher price. Early in our program, a number of knowledgeable American dealers told me the best goods we offered couldn't be Yogos because they had never seen a Yogo that good. It took a little doing to convince them that any sapphires but the light steely blues were Yogos.

Arnold Baron was the first American to achieve any success in the cutting and marketing of Yogo sapphires, yet his long-range plans were not to be fulfilled, for Baron would also be the first to realize that the geology of the Yogo dike was not, as Stephen Clabaugh had stated in United States Geological Survey Bulletin 983, "exceedingly simple." In his talk to the 1982 International Gemological Symposium, Baron told why his venture at Yogo was halted.

Although our cutting and marketing programs were successful, the mining was not. It proved far more costly than anticipated. The sapphire bearing dike has been traced for a distance of more than four miles in an essentially east-west direction. It is in a mountainous area difficult to reach and not easy to mine. In some areas the sapphire bearing matrix is virtually on the surface; in others there is considerable overburden. No one knows for sure how deep the dike runs. We know the British went down 300 feet and did not run out of dike. However, there are numerous faults in the dike, making the mining much more costly and complex.

Our geologist had worked with petroleum, not with gemstone mining, so he had to learn as he went. At the end of our lease period, we concluded that, if we had it to spend, which we didn't, we could easily spend millions finding out what we were doing wrong in the mining, and so with considerable reluctance we did not exercise our option to buy the property.

I have gone into considerable details concerning our experience with the Yogo, because the problems which we encountered will have to be solved by anyone who undertakes this enterprise.

When Baron did not exercise his option to purchase the property, Siskon, Inc., immediately put Yogo up for sale. The mine was not idle long, for a buyer soon came forward. With the sale of the mine in August, 1968, Yogo was back on the front page of the *Great Falls Tribune*.

FAMED SAPPHIRE MINE NEAR UTICA SOLD

LEWISTOWN—The famed Yogo sapphire mine near Utica in the Little Belt Mountains has been sold.

The office of H. E. Chessher, Sr., said he had entered into a purchase agreement with Herman Yaras of Los Angeles. Chessher is president of the Siskon Corporation of Reno, Nevada.

Chessher's office would not confirm the purchase price. A qualified Lewistown source said the mine was sold for $585,000, which includes all rights, mineral and surface. . . .

. . . The three men (company representatives) who visited the mine last week said the area will be subdivided into lots and sold to the general public.

Siskon, after ten years of involvement with the Yogo dike, much of which was spent in court, had probably broken even. The new owner, Sapphire Village, Inc., not only planned to mine and market sapphires, but, more importantly, to subdivide a part of the property into lots for sale to rockhounds. The plan was certainly one of innovation and imagination. It is history's loss that Charles Gadsden was not around to offer his comments when the plans were outlined in the March 30, 1969, *Great Falls Tribune*.

HISTORIC YOGO GULCH WILL AGAIN YIELD BRILLIANT BLUE SAPPHIRES

UTICA (AP)—Locked in an ore vein in a remote section of central Montana is one of the most valuable gem deposits in the United States—millions of dollars worth of brilliant blue sapphires.

A group of California men hopes it has the key to unlock the riches. . . .

Litigation over property rights tied up the land for years. The property is now owned by a Delaware corporation, Sapphire Village, Inc., licensed to do business in Montana and California.

President Herman A. Yaras from Oxnard, Calif., says the firm's principle shareholders live in Oxnard. Some of them, Yaras said, were shareholders in a corporation formed in Montana which unsuccessfully attempted to mine the area in 1964-65, Yaras told the Associated Press.

That company failed, Yaras said, because it had little knowledge of production and distribution methods.

Sapphire Village will begin mining the stones this summer and cut them on machines imported from Germany. The cutting operation, which will be done in Lewistown. . . .

By using modern mining equipment and automated cutting machines, Yaras said, the new company hopes to operate at a profit. All the stones, regardless of size,

> he said, will be used for jewelry purposes.
>
> During the early 1960s, rockhounds flooded into the mining area but they have been locked out for about three years. The new firm will let amateurs in on a limited basis only, Yaras said. To qualify for digging rights, amateurs must purchase homesite plots in Sapphire Village, a nearby area owned by the company.
>
> Yaras also says the Yogo stones retail for more than $200 per carat and are worth anywhere from $1 to $30 uncut. . . .

Sapphire Village, Inc., was going to generate a bit of cash flow by selling lots to rockhounds, hopefully to many of the same ones who had flocked to the dike a decade earlier when the property was unguarded, or those who had climbed fences or walked around roadblocks when it was posted. While a gem or mineral collector might be content with purchasing specimens, a true rockhound searched for and found his specimens. Sapphire Village was aimed directly at the hard core rockhound. The promotional literature was ready for mailing in spring, 1969.

> WORLD FAMOUS YOGO SAPPHIRES—WILL ONCE AGAIN BE AVAILABLE TO GEM BUYER AND COLLECTORS
>
> NOW—YOU CAN HUNT FOR FABULOUS BLUE SAPPHIRES IF YOU INVEST IN PROPERTY AT SAPPHIRE VILLAGE
>
> A UNIQUE OPPORTUNITY FOR THE GEM HOBBYIST AND COLLECTOR—VACATION PROPERTY WITH PERPETUAL ACCESS TO THE YOGO MINE
>
> . . . Because the rich deposits of sapphire still lie hidden beneath the earth the mines are again being readied for commercial mining. In addition, knowing of the intense interest of rockhounds and knowledgeable collectors, the owners are subdividing adjacent lands into 10,000 sq. ft. minimum lots up to ⅓ acre lots (approximately) giving the purchaser permanent rights in the deed to hunt for sapphires of high commercial quality and value along a minimum of one virgin (never before worked) mile of the sapphire bearing dike. . . .
>
> Lots runs from a minimum of approximately 10,000 square feet to ⅓ acre and will be priced from approximately $1,000 to $1,500, and there

> are a limited number of lots available. As soon as your money is in escrow, a deed will be issued permitting the "mining" of sapphire bearing material on the mine property under certain restrictions pertaining to hours of activity, the use of hand tools only and the hauling away of certain quantities of sifted material for sluicing at the lot owner's home or leisure.

Sapphire Village was certainly one of real estate's more unusual subdivisions and received quite a bit of publicity, especially in rockhound circles. Deposits were already rolling in when the April, 1969, *Lapidary Journal* carried an article titled "Precious Yogo Sapphires Will Again Gleam Among the World's Precious Jewels." The author was Sapphire Village president Herman Yaras. Written in breathless fashion, it was really thinly-disguised advertising for selling more lots.

> . . . The very uniqueness of the idea of throwing open to the public the right to acquire an actual portion of the ownership of building lots adjacent to the miles-long Yogo dike in Judith Basin, Montana, together with a right in perpetuity to dig sapphire-bearing material from a virgin portion of the dike especially set aside entirely for these parcel owners, has appealed to the imagination of many hundreds who have responded to the invitation of the owners of the sapphire dike itself and the incorporators of Sapphire Village, Inc., whose headquarters are located in nearby Lewistown, Montana. With the coming of the Spring thaws, can't you just imagine the excitement of all these people who have obtained these building lots and digging rights in exploring their very own portion of this fabulous sapphire-bearing dike?

Sapphires were probably the only thing that could have sold lots in the remote foothills of the Little Belt Mountains. Within one year, about fifty individuals or families had purchased lots on the subdivision, which was actually a part of the old Pagel Ranch—Charles Gadsden's "slum farm" sixty-four years earlier. The new residents of Sapphire Village had no complaints. They spent the summer days industriously shoveling, sifting and sorting their way through the daily limit of one hundred pounds of dirt. Many sapphires were recovered, including stones that weighed two or three carats in the rough. The daily one hundred-pound "mining" limit assured that no fortunes could be made, but was enough to keep Sapphire Villagers busy selling, grading, cutting, polishing, displaying and talking about their finds through the long Montana winter.

Sapphire Village, Inc., gave plenty of attention to promoting and selling real estate, but relatively little to the problems of commercially mining and marketing sapphires. The "modern mining equipment" and "automated cutting machines" never appeared and, as with every previous American mining effort, there was little operating capital. On its best days, the commercial mining operation washed about one hundred tons of dirt and recovered 1,000 carats. The gem quality stones, still without the benefit of a cutting and marketing system, were worth little. Four years after its founding, Sapphire Village, Inc., joined the growing graveyard of American companies that had failed at the Yogo dike.

This time there was no lengthy period of inactivity. One of the original Sapphire Village, Inc., partners, Chikara Kunisaki, of Oxnard, California, feeling compelled to protect his already substantial investment, bought the assets of Sapphire Village, Inc., and went from produce wholesaler to sapphire miner. Kunisaki's company was Sapphire International Corporation, which would be the first truly serious mining effort since the days of the English Mine. Yogo, and its new owner, were back in the *Great Falls Tribune* on July 29, 1973.

CALIFORNIA PRODUCE MAN REACTIVATES YOGO MINE

Yogo sapphires, known around the world for their consistent cornflower-blue color and brilliance under artificial light, are being extracted again from the Yogo Mine near Utica under new ownership.

When a proposed Sapphire Village subdivision near the mine failed to attract investors a few years ago and the bills came due, Chikara Kunisaki of Oxnard, California, one of the original partners, bought out the interests of his partners.

Kunisaki has 40 men employed at the mine on two shifts and is developing the gem-carrying vein first discovered in 1895 by gold placer miners. The $45,000 monthly payroll is a boost to the Judith Basin County economy. . . .

> Kunisaki, in a telephone interview, said, "I've been involved in this mine for about five years, since it originally started out as Sapphire Village, Inc., selling land to rockhounds and letting them work part of the vein with hand tools. I feel this is an extraordinary mine. We will promote tourism to it gradually, and we'd like to make it an extension of some regular tour, a side trip.
>
> "We have been struggling at this mine for the past four years. Production has been nominal in this development stage. We haven't been able to get into the good stuff yet. We feel the mother lode is down there and still hasn't been tapped. Eventually, it will lead to the betterment of the whole state. It has been a little hard to get good miners but we will have a good crew because it will be year-round work.
>
> The American-born Kunisaki, who raises celery and tomatoes and is involved in wholesaling produce, remarked, "I've had to learn this mining the hard way, nothing came easy. If we succeed it can only help the welfare of the State of Montana. This Montana stone actually is practically unknown in this country and doesn't get its rightful recognition.
>
> "Jewelers are concerned only with selling 'a sapphire.' The jewelry retailers are ignorant of how good a stone this is. We don't get the proper recognition for it but we'll get more. We haven't really made a dent in the market and we've got to overcome this. Yogo sapphire should be a household word. I think we'll be able to overcome 75 percent of this in time. . . .

But time was running out for Chikara Kunisaki who was sinking some $5 million into Yogo, most of it into a 3,000-foot-long tunnel driven eastward into the dike at the old American Mine site. The "Kunisaki Tunnel," as it became known, was quite modern; rechargeable electric locomotives pulled tiny, one-ton, rail-mounted ore cars, while miners used powerful pneumatic rock drills and blasted with light charges of dynamite. While mining methods and equipment were certainly adequate, geological knowledge of the dike was not. The tunnel encountered only small "pockets" of pure dike rock, not the massive dike formation that was anticipated. Chikara Kunisaki never had the opportunity to solve this geological mystery, for Yogo had taken him to the brink of personal bankruptcy. In fall, 1976, Sapphire International Corporation shut down all operations and joined Sapphire Village, Inc., in the lengthening list of American companies who came up short at the Yogo dike.

Continuing where Kunisaki left off came yet another company, Sapphire-Yogo Mines, Inc., headed by Victor di Suvero, who would be remembered as the most flamboyant of the would-be American sapphire entrepreneurs. Born in Turin, Italy, and a baron by heritage, di Suvero was raised in Tientsin, China, where his father commanded Italy's Far Eastern squadron in the years preceding World War II. After coming to the United States as a teenager, di Suvero pursued a varied career that included among other things, stints as a wallpaper salesman, art gallery manager and coal company president. In 1968, he successfully promoted a California jade mine, from which sapphire promotion seemed to be a logical and simple next step. On December 20, 1978, the Yogo sapphire deposit, and the plans Victor di Suvero had for it, earned a front page article in the *Great Falls Tribune*.

FERVOR OF OLD WEST PUSHES MODERN SAPPHIRE MINER

UTICA (NEA)—. . . After old Jake Hoover made his discovery 83 years ago, an English company moved in and mined the sapphires until 1929 when the diggings were closed because of production problems and legal snarls. They lay dormant for thirty years—until Victor di Suvero came along.

The place today is bucolic enough to have 400 head of elk drifting through its rock wall canyons. But for di Suvero, it represents the biggest bonanza in the life of a mercurial entrepreneur whose only prospecting tools are leases, geological

> surveys and speculative capital.
>
> Since January, 1977, di Suvero's Sapphire-Yogo Mines, Inc., has held a lease in perpetuity on the central Montana vein and had already begun mining operations.
>
> "We're going to have an impact on the sapphire market of the world," says di Suvero in a burst of enthusiasm. "We hope to get $40 million worth of sapphires out of the mines annually. Less than five percent of the deposit has been mined. . . ."
>
> In the fall of 1968, di Suvero had revived an old jade mine in California's Mariposa County. A mining engineer at Yogo Gulch heard of him and asked, "Would you like to do sapphires next?"
>
> For most of the next decade, di Suvero tried to assist the company that owned the rights to structure and finance a viable program, while $5 million went down the drain. He finally moved to take over the entire mine himself a couple of years ago. It required a total commitment and di Suvero estimates he personally has sunk $1.4 million into the venture. . . .

As a somewhat successful promoter, di Suvero directed his attention to matters where his expertise would be most effective and which had been virtually ignored by previous companies—advertising and marketing. For marketing and distribution of Yogo sapphires, he established a San Francisco company, Sapphire Trading Co., Inc. Di Suvero's approach to advertising Yogos was imaginative, creative and aggressive—so much so that *Jewelers' Circular/Keystone*, a leading trade publication, gave the ad campaign a full page in a 1978 issue, the first official trade recognition of the Yogo sapphire.

> **THE $1 MILLION SAPPHIRE CAMPAIGN**
>
> This December, when millions of men who read *Esquire* and *Sports Illustrated* are told to give diamonds, millions of women who read *Vogue* and *Cosmopolitan* will be told to desire sapphires.

Montana sapphires at that.

It's the kickoff to an ambitious $1 million advertising campaign designed to make Montana sapphires the premiere American gemstone and run by Sapphire Trading Co. This recently-formed San Francisco firm is a mini-De Beers that cuts and markets the mining output from the recently reactivated Yogo Sapphire Mine, Lewistown, Montana. The Yogo mine is slated to produce 120,000 carats of cut sapphires in 1979, and at least 200,000 carats of cut gems annually when the year-round underground mining begins in 1981. Reserves are said to be great enough to sustain the latter production figure for a minumum of 50 years.

"We know we've got the supply of sapphire. Now we're going to see if we've got the demand," sas Victor di Suvero, president of Sapphire Trading Co.

To create that demand, Yogo is launching three print advertising campaigns, each geared to a different kind of magazine and audience.

The first series of ads will appear in "location" magazines, 11 inflight and 9 city publications (among them *New York, Philadelphia, Phoenix* and *Texas Monthly*). Ads will carry an umbrella theme, "Sapphire, the American gem," and show Yogo jewelry in a variety of places (including Rome, London, Paris, Singapore) and settings (the ball park, race track, theater and supermarket). According to di Suvero, inflight magazines are read by nearly 20 million people monthly and "anything in them is taken for granted as being established."

Montana sapphires also will be seen in a second promotion designed for women's magazines such as *Vogue* and *Cosmopolitan*. Ads in these books will feature what di Suvero calls "discreet nudity." Translated that means naked models wearing only Yogo jewelry shown in the gauziest of soft focus. The catchy tag line for these ads is: "If you've got sapphires, you don't need anything else."

Di Suvero expects a lot of mileage from these particular ads. Already, he tells *JC-K, Penthouse* has sent its cameras to snap Yogo's cameras photographing these ads and the lovelies who will adorn them. The end result will probably be a four-page color story early next year.

Of most interest to jewelers will be di Suvero's third ad campaign. In this promotion, Yogo will buck heads with De Beers and try to dethrone diamonds as the mainstay jewelry gem. Ads will show, in di Suvero's words, a "John Travolta-type with a gorgeous 15-year-old blonde model in a *Saturday Night Fever* locale." Copy will read: "Diamonds go to the country club; sapphires go to the disco."

In other words, di Suvero explains, "sapphires will be pitched as the with-it, good-time gem."

These anti-diamond ads will first appear in youth-oriented magazines like *Seventeen* because, as di Suvero says, "the young are not yet brainwashed about diamonds."

In the end, di Suvero hopes to turn on every segment of American influence "from tired businessmen to Pepsi generation disco types" to Montana sapphires. He admits others have tried to do the same thing without success, but says none were willing to spend the money he is. Further, he feels that right now is the first time Montana sapphires have a chance to take off as a gem in their own right.

"Diamond prices are much too high," di Suvero says. "So we're telling Americans of all ages and life styles that we've got a real alternative to diamonds—one that is reasonably priced, beautiful, of lasting value and a product of their own country. We think that's an unbeatable combination of factors."

So does Communimark, Yogo's ad agency. In fact, the firm is so convinced Montana sapphires will catch on, that it is accepting payment of its $1 million fee in the form of Yogos.

"It's gem investment at its best," di Suvero says. "They're putting our stones in a vault, waiting for their ads to do the job of selling them, then cashing in."

The *Jewelers' Circular/Keystone* article was probably the best promotion di Suvero ever managed, for the ads themselves were never run. Di Suvero's talents seemed suited only to advertising, for he virtually ignored the more basic matters of geology and mining. Also, he had been underfinanced from the beginning; much of his effort had been expended in raising capital to meet the next lease payment. Di Suvero's advertising-oriented venture at Yogo ended abruptly in fall, 1979, when he was unable to make the lease payments.

Di Suvero, in his effort to create a marketable image for the Yogo sapphire, had even supported a move to have the Yogo declared the "national" gemstone of the United States. In the late 1970s, the American Gem Society was assembling a collection of jewelry mined and made entirely in the United States for presentation to the Smithsonian Institution. The renewed interest in native gemstones prompted an interesting debate over the qualities a stone should have to qualify for such an honor. John Sinkankas, the noted author and gem and mineral expert, suggested some ground rules for selection. Sinkankas thought the national gemstone might be one found exclusively in the United States, such as benitoite, the beautiful blue crystals of barium titanium silicate that occur only in California; or it might be a stone of broad commercial appeal, such as the ornamental jaspers and agates that are sold to tourists by the ton; or perhaps it should be at least a semi-precious stone, such as the California pink or Maine green tourmalines. Finally, Sinkankas suggested that perhaps the national gemstone should be one of the four classic and truly precious gemstones—diamond, emerald, sapphire or ruby.

Not surprisingly, the debate followed commercial lines. Southwestern stone dealers thought turquoise should be chosen, for it was first mined and worked by Native Americans. Tourmaline dealers believed their stone was the only logical choice; tourmaline, symbolic of the national polyglot population, occurred in many colors and crystal shapes and, furthermore, was mined on both coasts. Di Suvero's Sapphire-Yogo Mines, Inc., argued that Americans, as a proud people, deserved better than a relatively unknown semi-precious stone for their national gemstone; without question, they would demand one of the four classic and truly precious gemstones. And, of course, the only such precious gemstone of commercial significance in the United States was the Yogo sapphire.

The national gemstone debate was never carried to its conclusion. If it had been, the Yogo sapphire would probably not have been selected. The big mark against the Yogo was that it came from a single source, which was controlled by a single company. Cornelius Hurlbut, author and professor emeritus of mineralogy at Harvard University, summed up the problem, saying, "To declare the Yogo our national gemstone would be to give free advertising to one mining outfit. We've got to watch out that we don't end up promoting a company instead of a gemstone."

It would be an understatement to say that, through 1979, the American experience at the Yogo dike had been less than successful. The only Americans to walk away from the Yogo dike with a little more money than they had put into it had been John Burke and Pat Sweeney, the original American Mine claim holders, and S. S. Hobson and Matt Dunn, partners in the first New Mine Sapphire Syndicate, all of whom profited through sale of their properties. At least thirteen separate American efforts to emulate the success of the English Mine had failed at the business of mining and marketing Yogo sapphires. Those who managed to at least break even did so only through resale of the property.

What had gone wrong? One of the problems was a basic American inability to comprehend the complexities of precious gem mining and marketing. Most Americans still equated sapphires with gold, particularly with the frontier tales of picks, shovels and luck that often ended happily with a pouch of raw nuggets. Those gold nuggets, which needed neither cutting nor marketing, were as good as—even better than—money. The American companies that worked at Yogo at the turn of the century and even as late as the 1960s still considered sapphires to be "blue gold." They rushed to the Yogo dike with a pickup truck full of shovels, dug and sweated through the Montana summer, and succeeded in reaching their shortsighted goal. At summer's end, all could display a bag full of rough Yogo sapphires. Only then did they realize that rough sapphires, unlike raw gold nuggets, had relatively little value. A great deal more work and expertise was needed to turn them into true "blue gold."

Americans saw the English Mine as a model of success. But even that was a simplistic view. All the Americans ever saw was a stubborn Englishman and his work crews shoveling dirt through sluices and reports of fat earnings and profits. Those reports, together with the exciting idea of "let's go dig some more" formed the heart of the Yogo Sapphire Mining Corporation's stock promotion literature in 1949. But the Americans did not understand what happened to the rough Yogos after they reached London. They knew nothing of a system of gem cutters in four European countries or of the enormous advantages afforded by the established marketing network and promotional resources of Johnson, Walker and Tolhurst, Ltd. The inability to think beyond a bag full of rough sapphires was a primary reason for most American companies to rush to the Yogo dike dreadfully underfinanced.

Another reason for the American problems at Yogo was their peculiar interpretation of, and utter faith in, United States Geological Survey Bulletin 983. Stephen Clabaugh's report on Yogo was never intended to be the definitive geological investigation of the dike, but merely part of a general survey of Montana's corundum deposits. But the American absolute acceptance of United States Geological Survey Bulletin 983 as gospel had discouraged subsequent, more detailed geological investigation of the dike. If Clabaugh had erred, it was only in stating the geology of the dike to be "exceedingly simple." The work of Arnold Baron and Chikara Kunisaki had now proved otherwise. But in his estimates of reserves and overall future mining potential, Clabaugh would, in time, be proven remarkably perceptive and accurate. The trouble was in interpretation of Clabaugh's report; each time he had written the word "sapphire," most American promoters and would-be miners read only the word "gold."

A century earlier, many Montana miners had turned their backs on Montana sapphires because they weren't gold. Now, Americans were unable to strip away that enticing, simple frontier image of gold to consider sapphires as they really were—gemstones. None of the companies that failed to commercialize the had balance; all were limited in scope, interest and expertise, concentrating only on a few aspects of what was a very complex business. Many thought only of mining and nothing of marketing. Sapphire International Corporation, while creating a model underground mine, had not undertaken any geological exploration and investigation. Victor di Suvero had formulated an exciting advertising program, while Arnold Baron organized an effective cutting and marketing network; neither had given attention to mining or geology.

Since the Yogo dike was returned to American ownership, at least nine companies had spent some $10 million in equipment, labor, advertising, stock promotion, mining and time—to say nothing of the emotional cost in shattered hopes and dreams—only to learn that sapphires weren't gold, but an infinitely more complex mineral commodity. That lesson was their only reward, for relatively few sapphires had been mined from the Yogo dike since the closure of the English Mine. But, in that time, the millions of sapphires that lay locked in the dike had gathered a great deal of financial interest. In 1900, cut sapphires were worth about $30 per carat; during the 1920s, better stones were worth several hundred dollars per carat. By 1979, when Victor di Suvero was giving up at Yogo, fine cut sapphires, of the quality and color that characterized Yogos, were rivaling the value of fine diamonds.

The Yogo dike was indeed "the most important gem locality in the United States." It was an enormously rich mineral treasure, but not one that would be unlocked by one-sided efforts or the old frontier approach of a couple of shovels and a little luck. The Yogo dike would demand far more.

Chapter 5

The Return of the Yogo

After Victor di Suvero relinquished the Yogo lease in late 1979, Chikara Kunisaki put the property up for sale. To recoup his multimillion dollar investment in the dike, the price was set at $6 million. Considering the fact that not one American effort had been operationally successful at Yogo, it would not have been surprising if no market interest was shown. What happened was perhaps more surprising. Not one, but four individuals or groups appeared to consider the purchase of the Yogo dike.

Among those interested was Harry C. Bullock, a mining engineer who had studied at the University of Wyoming and the Colorado School of Mines. Bullock already had twenty years of experience in mining and heavy construction when one of his companies was contracted by Sapphire International Corp. to further develop the underground workings at the old American Mine site. In 1977, when Bullock Exploration, Inc., was working for Victor di Suvero, Harry Bullock brought in a consulting geological engineer and gemologist, Delmer L. Brown. But di Suvero was concerned solely with advertising and promotion, not mining and geology. When Sapphire-Yogo Mines, Inc., gave up in late 1979, Harry Bullock and Delmer Brown, together, took a long, hard look at the Yogo dike, not as stock promoters, jewelers, rockhounds, real estate developers or advertising men, but from the decidedly unromantic perspective of mining engineers and geologists. They saw neither "blue gold" nor a get-rich-quick opportunity, but a mineral resource which, if systematically exploited, could be profitable. Bullock then went to his friend J. R. Edington, a Spokane financier, to explain the potential of the property and the fact that it was immediately available. Edington was interested. "If we can get our hands on Yogo," the financier said, "I'm sure we can do something with it." Edington structured a limited partnership, American Yogo Sapphire, Ltd. In April, 1980, Harry Bullock took control of Yogo in the name of the partnership. Agreeing to pay $6 million for the property, American Yogo Sapphire, Ltd., became the fourteenth American company to try its luck at Yogo.

American Yogo Sapphire, Ltd., embarked upon the most ambitious funding program ever at Yogo—$7.2 million. Earlier companies had failed to raise far less, but American Yogo Sapphire, Ltd., would succeed largely because of fortuitous economic timing. The 1970s had brought rampant inflation and dramatic reorganization of the world economy. The basic cause was skyrocketing crude oil prices which eroded the value of most currencies and sent gold on an unprecedented upward spiral. Long fixed at $35 per ounce, the price of gold reached $100 in 1974; in a world of shaky paper currencies, the yellow metal became one of the most demanded investment commodities. Fine precious gems—diamonds, emeralds, sapphires and rubies—had nearly as much investment appeal as gold. As a store of wealth, precious gems were extremely portable; great purchasing power was represented in only a few easily concealed and transported carats. Although not quite as liquid as gold, gem values were universally recognized. And because of the inherent rarity of fine, natural stones, that value was assured. As demand for investment gems increased, their prices followed suit. In 1979, the Iranian revolution triggered another energy "crisis," creating still more inflation. By January, 1980, the dollar had hit a record low, while gold reached an all time high at a price of over $800 per ounce. Simultaneously, the prices of precious gems reached their own record highs.

All precious gems soared in price, with the greatest increases registered by the colored gems—emerald, sapphire and ruby. For decades, diamonds had dominated the precious gem markets, thanks to saturation advertising by De Beers Consolidated Mines, Ltd., which had conditioned, or "brainwashed," the consumer. Only diamonds, De Beers had decreed, were "forever." To young engaged couples, the diamond was the traditional—even mandatory—seal for the relationship. But, by 1974, diamond advertising had reached overkill proportions. At the same time, prices of finer diamonds began rising beyond the reach of the mass market. In seeking alternatives, consumers turned to sapphires, emeralds and rubies, stones that had hardly dented their potential markets. By 1980, colored gems such as sapphires, *and their sources*, were suddenly a timely and "hot" investment. American Yogo Sapphire, Ltd., was selling the right thing at the right time. In only eighteen months, the offering was sold out. The partnership had raised $7.2 million by October, 1981.

Bullock's original plan was to mine the Yogo sapphire roughs, cut them, and sell the gems to gem merchants and jewelry manufacturers, the same basic idea as put forth by his now-defunct predecessors. In late 1981, however, the results of a thorough cost analysis had raised serious questions; on paper, mining, cutting and selling unmounted gems would be, at best, a break-even propositon. In Bullock's view, that was not a fair shake of the dice for the investors who had put up $7.2 million. J. R. Edington suggested a solution: Take the operation one step further to the manufacture of sapphire jewelery, thus retaining the substantial price markup between cut gems and finished jewelry. This concept of vertical integration would permit the company to take the Yogo sapphires right from the rough to mounted gems in the showcases of retail jewelry outlets. There was no trade precedent to follow; if successful, this would be the world's only fully integrated precious gem company. Bullock recalls the change in plans was not greeted with unanimous enthusiasm. At one meeting of the limited partners, a woman investor complained loudly, "When I invested in this company, you told me we were going to mine sapphires. Now, all of a sudden, you tell me we're going to sell jewelry!"

Vertical integration could not only increase the profit potential, but could establish a unique market "identity" for the Yogo sapphires. Virtually all sapphires, regardless of origin, were cut in Thailand, then channeled into the world gem markets. By the time these stones reached the retail jewelry counters, they were simply "sapphires," blue corundum gems without any particular origin, history or story. Although a gemologist, through microscopic examination, might determine origin, to the consumer, a sapphire was a sapphire. But vertical integration would control a sapphire from mine to retail store, thus making it possible to market a different sapphire, one with a true identity.

While Harry Bullock and J. R. Edington hammered out the idea of vertical integration, Delmer Brown was piecing together the true story of the Yogo dike. He had already amassed an extensive collection of historical records, letters and photographs dating back to the English Mine; more importantly, he had begun the most thorough geological investigation of the Yogo dike ever undertaken, one that did not rely on United States Geological Survey Bulletin 983 as a base.

All of the early geologists to visit the dike had expressed opinions of its true dimensions and structure. Most believed the dike went "clear down"; some others suggested just the opposite, that most of the dike had eroded away to leave only the "roots" of the original emplacement. The question of depth was not merely academic, for the economic value of the deposit was dependent upon the actual volume of sapphire-bearing rock present. Brown quickly showed that the erosion of the dike was both minimal and extremely recent. In places along the dike wall exposed by mining, he could put his finger on the contact point between the uppermost part of the Madison limestone formation and the overlying shale strata—the precise point where the ascent of the magma had been halted fifty million years ago.

Other evidence showed that erosion had just brushed the tip of the dike. The greatest erosion had occurred where Yogo Creek had cut a 200-foot-deep canyon through the dike. If only the "roots" of the original dike emplacement remained, enormous quantities of sapphires would have been carried far downstream in the Judith River. But the alluvial sapphires had never even reached the Judith; in fact, none had been found more than a mile below the dike in the Yogo Creek gravels. Therefore, the discoveries of Jake Hoover and Jim Ettien had been made at the earliest possible geological time. If surface erosion had proceeded any slower, Hoover would never have panned his blue pebbles and Ettien could never have noticed that faint depression transecting the bench lands. Only nature would know of the existence of the Yogo dike.

Based upon United States Geological Survey Bulletin 983, most would-be sapphire miners had pictured the dike as a textbook model—a narrow magmatic intrusion of "pure" dike rock neatly lying between two limestone walls. After study of the surface and the underground sections accessible through the American Mine tunnel, Delmer Brown concluded that the dike geology was quite complex. Geological alteration had occurred within the original fault before the magma ever rose to fill it. The fault had served as a natural watercourse within the Madison limestone, and portions of the limestone walls had dissolved into caves and cavities. Most of these eventually collapsed and filled with breccia, a physical admixture of limestone and other rocks, usually loose, but occasionally cemented weakly together.

As the magma ascended, it filled these brecciated cavities, solidifying into what Brown called "pre-dike" breccia. Alteration continued after the emplaçement of the dike as groundwater created new cavities in the limestone walls adjacent to the solidified dike rock. Later, earthquakes and intense ground movement associated with new faulting collapsed these cavities, filling them with a breccia composed of fragments of limestone *and* dike rock. Brown identified these as "post-dike" breccia. Differentiation between the two types of dike-related breccias was not difficult. In pre-dike breccia, the magma had flowed between the fragments, solidifying to form the matrix of a well-cemented breccia. The post-dike breccia, however, was a mix of limestone, dike rock and even pieces of the overlying shales, all weakly cemented together by the action of water solutions.

In many places, the dike still existed in textbook simplicity; "pure" dike rock terminated abruptly at a sharp contact plane with intact, smooth limestone walls. In other areas, entire sections of the dike had collapsed into huge pockets of post-dike breccia, containing a great deal of limestone but relatively little dike rock. The sapphire content and, thus, the economic value of these areas was hardly significant. The American Mine tunnel, unfortunately, was driven through large areas of post-dike breccia in a futile search for a formation of "pure" dike rock that no longer existed.

Early mining had shown the dike existed at least to a depth of 300 feet. Later, core drilling extended this to 500 feet. Through careful study of the dike rock, Delmer Brown found evidence that the dike existed to a far greater depth. The dike rock contained fragments of all the other geological formations beneath the Madison limestone. Most significant were many pieces of the Precambrian basement complex that lay 7,000 feet beneath the Madison Formation—pieces that could only have been carried upward on the ascending surge of magma. Those early geologists who guessed the dike extended "clear down" were right; by geological inference, the Yogo dike extended *at least to a depth of 7,000 feet.*

Delmer Brown then began connecting the creation of the dike with the nature of its sapphires. Yogos were a very unique and distinctive sapphire. Unlike others that exhibited every color of the spectrum, Yogos occurred only within a tight range of blues. Also unlike other sapphires, Yogos had no color zonation whatever and only rare inclusions. Clearly, Yogos had been created under very unusual conditions.

The zonation, inclusions and broad color range of foreign sapphires indicated crystallization from their molten components at relatively shallow depths, in conditions of low temperature and pressure and over a short period of time. Such conditions of chemical and physical disequilibrium would account for their many imperfections; color could not be evenly distributed and the rapid crystallization would "trap" many mineral inclusions foreign to the basic crystal structure.

Conversely, the Yogo characteristics indicated crystallization in extremely high temperatures and pressures and over a relatively long period of time—conditions of chemical and physical equilibrium possible only at great depth. The iron and titanium chromophores were evenly distributed throughout the slowly developing aluminum oxide crystal lattice. The absence of inclusions also testified to a lengthy period of crystallization, in which each slowly developing crystal could literally purge itself of foreign minerals, rather than trapping them within. There was even tangible evidence that Yogo sapphires had crystallized at great depth. Microscopic examination of sapphires still imbedded in the dike rock matrix revealed physical etching on their natural crystal faces—marks that had to be created *before* solidification of the magma. The Yogos had obviously been formed at great depth, then carried upward thousands of feet—perhaps even several miles—by the slowly thickening magma, being scraped and etched along their journey.

Even the size and shape of the Yogos—somewhat smaller and flatter than other sapphires—seemed to confirm this theory of creation. As the Yogos were carried upward on the stream of magma, they were subjected to a certain degree of gravitational separation.

Since sapphires are considerably heavier than the magma, only the smaller and flatter stones would logically have sufficient buoyancy to be carried to the top of the dike. Larger and rounder stones, therefore, could be expected to be found deeper within the dike. And English Mine records show that the largest sapphires recovered came from the deepest workings.

American mining interest at Yogo had traditionally focused on "richness," specifically, the number of carats found in one ton of dike rock, a concern that doubtlessly stemmed from the "ounces per ton" figures used to describe the richness of gold and silver hardrock ores. To American miners, the "average ounces per ton" figure best indicated the economic potential of a precious metal deposit. At Yogo, the American miners demanded a similar "carats per ton" figure and found one in United States Geological Survey Bulletin 983. From study of incomplete English Mine production records, Stephen Clabaugh had noted that "the average yield was between 20 and 50 carats per ton." American miners "refined" this broad estimate to a simpler "average" richness of 30 carats per ton. Investors, and even some miners, expected that for every ton of dike rock mined and washed, thirty carats of sapphires would be recovered. Mine production could thus be conveniently projected; if 100 tons of dike rock were washed, the recovery would be about 3,000 carats. Some of the announced (but highly unlikely) production figures from the 1950s can be traced directly to "average richness" figures.

After a detailed computer study of English Mine production records and his own sample data, Delmer Brown concluded that the term "average" did not realistically apply to the sapphire content of the Yogo dike. The actual sapphire content varied dramatically within the dike, ranging from only five carats per ton to a high of fifty carats per ton. Even more importantly, Brown found that the size of the sapphire crystals varied inversely with their concentration. This meant that the sapphires in 40-carat-per-ton dike rock were smaller than those in 10-carat-per-ton dike rock. The popular "average richness" figures were, therefore, quite misleading, since one ton of 5-carat-per-ton dike rock could be many, many times more valuable than the "richer" grades.

The areas of pre-dike and post-dike breccias further complicated any estimates of richness. Most of the post-dike breccia was not worth the effort to mine it. Some of the pre-dike breccia, however, seemed to have acted as a natural "trap," literally straining out the sapphires as the magma surged through the breccia fragments. The sapphire content of some pre-dike breccia can easily exceed that of the richest dike rock. "Average richness" was never the key to successful mining; the key lay in selective mining based upon detailed geological mapping of the dike. As a result of Delmer Brown's geological investigations, the American Mine tunnel was permanently closed and future mining operations directed to a surface location near the site of the English Mine.

Geological studies and surveys were important, but some lessons were learned by trial and error. Harry Bullock's early mining operations in 1980 and 1981 were limited and somewhat experimental. In the hope of accelerating the weathering process, a bulldozer was run over the newly mined rock to physically break it down. Many sapphires in the jig concentrates soon appeared with clean, unetched, conchoidal fracture surfaces. Even in the soft earth, the weight of the bulldozer was shattering about ten percent of the sapphires. The bulldozer was immediately assigned to other tasks and weathering, as it had been since the English Mine days, was left to the elements.

The next problem faced by Bullock was cutting. With United States mining costs at Yogo relatively high compared with those of foreign mines, cutting costs had to be minimized if Yogos were to be competitively priced. The cutting alternatives varied widely in geographic location, cost and quality. The finest cutting could be conveniently done on automated cutting machines in the United States. While the cuts would be technically perfect, the cost of $20 per stone would price the smaller Yogos right out of the market. Foreign cutting centers were located in Germany, Hong Kong, Thailand, India and Korea. Visiting cutters in the Far East, Delmer Brown found the lowest costs in India and Korea—and the lowest quality. He also found among the Indian and Korean cutters an unwillingness to adjust their standards upward. In Bangkok, Thailand, he found low costs but, more importantly, an interest in learning to cut to improved standards. At forty cents per stone, Thai cutting was fifty times cheaper than that in the United States. But the philosophy behind Thai cutting was to cut for maximum weight retention, even if that meant compromising the brilliance of the finished gem. Brilliance in any cut stone depends upon proper geometric proportions and angles that maximize the internal reflections of the light striking the stone. But in oriental markets, a gem's value is primarily determined by carat weight.

Delmer Brown offered hundreds of Thai cutters the opportunity to demonstrate their skills. By the end of his second trip to Bangkok, he had assembled a small network of cutters whom he had "educated" to meet his standards and who began the first production cutting of the Yogos. During the next two years, Brown would make another ten trips to Thailand, spending six months in Bangkok checking, teaching and expanding his network of cutters.

Some gem merchants have described Thai cutting circles as an "adventure," for some Thai cutters place profit above morals or ethics. Gems have been known to simply disappear; cutters explain their loss as breakage or claim the presence of flaws prevented cutting to full weight. Many "losses," of course, go into the cutters' pockets. Some Thai cutters have even been known to substitute inexpensive synthetic sapphires and rubies for the natural gemstones they were paid to cut.

A security system was imperative, and Delmer Brown helped establish one that began at the mine itself. His computerized geological data permitted accurate projection of mine recoveries, both in size and number of sapphires, depending, of course, in exactly which part of the dike the particular run of dike rock had originated. All the recovered sapphires were then shipped to the company offices in Aurora, Colorado, near Denver, for sorting and grading. First, the mine production was sorted into culls and "cuttables." The culls, stones that were too small, too flat, or otherwise unsuitable for cutting into gems, were sealed in packets that were weighed, numbered and locked in a bank vault. That assured that the culls would never enter the gem trade as competition, nor be available for possible substitution as a cuttable. The cuttables were then carefully classified by a complex system of color, weight, quality and shape. There was even a separate classification for the two percent of all Yogos that were violet or purple.

Small numbers of a particular classification were then sealed into coded packets, the contents of which had been counted, weighed and recorded. The packets were then shipped to cutters in Bangkok. Within a very small margin of error, Brown knew the number, size and shape of the gems that would be returned within each coded packet. Each gem was then examined microscopically to assure there had been no substitution. Since no other natural sapphire could duplicate a Yogo, the primary concern was guarding against possible substitution with synthetics.

The bigger Yogos, in which cutting costs were not a factor in market pricing, were sent to United States cutters. German cutters were used for specialty cuts, and a smaller number of intermediate-sized stones were sent to Hong Kong cutters. The majority of Yogos, however, were cut in Bangkok, where cutting quality was constantly being improved.

The first two years of American Yogo Sapphire, Ltd., had been devoted to geological and mining studies, finding cutters, procurement of capital and general planning. By 1982, attention began shifting to matters of administration and marketing and progress became more apparent. To handle organizational and administrative concerns of the growing company, Harry Bullock invited a friend, Dennis K. Brown, to accept the position of president. Dennis Brown's professional experience included ten years with Touche, Ross & Co., a leading accounting and business management firm. When he joined the company in January, 1982, Dennis Brown recalls "there was little more than a hole in the ground and a couple of dollars in the bank." Actually, there was over $2 million in the bank, but that was hardly enough to cover the ambitious plans for commercial mining, cutting, advertising and marketing. Brown used his organizational and administrative skills to write an excellent company business plan, instituted many efficient inventory and accounting procedures, and established all of the major corporate systems, from cutting and manufacturing right through to distribution. One of his biggest jobs was to assure that American Yogo Sapphire, Ltd., unlike its predecessors, would not be forced to sell stock on the Lewistown street corners or push lots to rockhounds. Together, Bullock and Brown pursued lengthy negotiations with Citibank that did not conclude until a group of New York financiers was given a tour of the mine. A $3 million line of credit was secured with Citibank and quickly renegotiated upward to $5 million. Big business, at last, had come to Yogo.

Big business also called for a name more commensurate with the future image the company was hoping to build. American Yogo Sapphire may have sufficed for turn-of-the century gem marketing, but it would not do for the market of the 1980s. Bullock conducted a survey among his limited partners and associates to find a more suitable corporate name. Of dozens submitted, the one chosen was *Intergem*. By spring, 1982, American Yogo Sapphire, Ltd., had formally become Intergem, Ltd.

Just as "Yogo" had disappeared from the company name, so, too, would it be removed from the coming line of sapphire jewelry. Yogo, for all its history and frontier homeliness, was not the name upon which to build a modern marketing image. To establish an identity, the name "American" or "Montana" was necessary. "American" was chosen for its obviously broader market appeal. The name "American Sapphire" still needed the right qualifier, an adjective that would convey to the consumer the richness of quality of the sapphire they would purchase. That word had already been employed seventy years earlier by Johnson, Walker and Tolhurst, Ltd., in their promotional booklet, *A Royal Gem*. By summer, 1982, the venerable Yogo had been resurrected by Intergem as "The Royal American Sapphire."™

The first line of Royal American Sapphire jewelry contained only thirty designs and appeared in late summer, 1982. Designs were chosen that would best display the beauty and brilliance of the sapphire; gold mounts were selected to best complement the rich colors. The gold jewelry was manufactured in New York City, then shipped to Colorado where the sapphires were mounted. As test marketing began, the existence of Intergem and the reappearance of the Yogo sapphire were acknowledged in an article in *Jeweler/Lapidary Business* that concluded with this encouraging comment:

> The awesome, centuries-old fascination with sapphires is boldly rekindled with this magnificent gemstone. This special sapphire had represented an irresistible prize for the treasure hunter for many, many years. It appears now, that the beautiful blue sapphires of Yogo will finally take their rightful place with the jewelry customer because of the modern application of mining and cutting techniques and well-thought-out marketing approach of well-designed and affordable jewelry.

By the end of 1982, twenty-five dealers in four western states were carrying the Royal American Sapphire line. By the end of the year of test marketing, wholesale-level sales had amounted to $200,000. Intergem, with its multi-million dollar investment in the Yogo dike and a costly and risky commitment to the concept of vertical integration, clearly had a long way to go. Several months of test marketing, while not particularly impressive in the financial columns, was encouraging enough to move Intergem toward a national marketing program for 1983.

Another Intergem executive position was filled in early 1983 when Steven M. Droullard was named vice president for marketing. Droullard's background included fifteen years of experience in gem cutting, goldsmithing and jewelry making, as well as previous experience with Yogo. In the late 1970s, he had served as marketing manager with Sapphire-Yogo Mines, Inc., and in fall, 1982, as a sales representative with Intergem. Droullard began formulating a program that would attain a viable national marketing position through a system of authorized dealers.

Marketing the Royal American Sapphire demanded a unique approach, one that would offset its slightly higher price necessitated by the higher American mining costs. It was essential that the identity of the stone be established and maintained. The Royal American Sapphire could not be sold as just any sapphire, but only as a very special sapphire, one with its own distinctive origin and story. To achieve this, Intergem would have to maintain full control over all advertising, and jewelers would be permitted to sell the stones only under the name of the Royal American Sapphire. Steve Droullard's marketing program was based on the principle of sales incentive. To qualify as an authorized dealer, a jeweler would have to reach a specified sales level. In return, he would qualify for Intergem's prepared newspaper, television and radio advertising material, an advertising allowance applied to the purchase of new stock, and an array of attractive in-store display materials.

The reappearance of the Yogo sapphire on the American jewelry market brought some interesting reaction. Many jewelers found their American origin appealing; others, ironically, found a problem in the unusual beauty of the stones. They were disturbed by the radical difference of the Yogos when compared side by side with the foreign sapphires they were accustomed to selling. Claiming the Yogos made their foreign sapphire stock "look bad," some dealers asked Intergem's sales representatives "to stop back after I've moved my existing stock."

Although the striking beauty of the Yogos created some initial resistance, it soon became the biggest selling point. In New York City, Steve Droullard persuaded Saks Fifth Avenue jewelery buyers to make a side by side comparison of sapphire jewlery. Saks invited sapphire suppliers, including Intergem, to present their best pieces. After the comparison, Saks Fifth Avenue became one of the first chains to carry the Royal American Sapphire line.

Since the early days of the English Mine, Yogo sapphires had borne the stigma of being "too small," an image created a least in part by the flagrant marketing misrepresentations of origin, when the bigger Yogos were sold as oriental sapphires. But now, in the 1980s, the "smaller" Yogos were just what the market demanded. Doubtlessly, the most admired sapphire of the 1980s was the magnificent nine-carat stone set in the engagement ring that England's Prince Charles presented to Lady Diana Spencer, a stone that focused enormous public attention on the beauty and new stylishness of sapphires. Such sapphires, however, were precisely what the millions of individuals who make up the mass gem market could not hope to afford. After precious gem prices soared to record highs in 1980, the United States economy fell into a major business recession in 1982. As money became tight, demand for the larger precious gems lessened. Consumers began turning to smaller, more affordable gems—precisely what Intergem was marketing with the Royal American Sapphire line. Once again, economic timing was on the side of Intergem.

While Intergem struggled to gain a foothold in the jewelry market, it enjoyed little public attention and only the most perfunctory acknowledgement by the jewelry trade. But that was to change quickly, for the Royal American Sapphire had one final attribute, the publicizing of which would hit the gem trade like a bomb. So great was the expected controversy that Intergem had purposely delayed publicizing this point until it had achieved a stronger market position. By summer, 1983, this ultimate statement of the quality of the Royal American Sapphire was already causing some uneasy stirrings in the gem trade. In an informal discussion with another gem dealer, Harry Bullock admitted that this point would become the focus of Intergem's future advertising. Aghast, the dealer looked up and replied, "You don't mean you're going to bring *that* up. You can't—you'll destroy the industry!"

That first advertisement, which was destined to trigger lasting controversy and bitter debate within the gem trade, appeared in trade journals in late summer, 1983. Aimed at selling the Royal

American Sapphire to retail jewelers, the opening words were:

BE AN EXCLUSIVE DEALER OF THE WORLD'S ONLY
GUARANTEED UNTREATED SAPPHIRE

From that time on, every Royal American Sapphire sold would be accompanied by a "Certificate of Guarantee," they key words of which were:

> Unlike most sapphires sold today, the Royal American Sapphire is not heat treated to improve its color.

The words were simple, but they rocked the foundations of the gem industry. Intergem was daring to market the Yogo dike sapphires with a written guarantee that every single stone was untreated and completely natural—a claim no other sapphire source in the world could make. Intergem had succeeded in creating a distinctive image for the old Yogo sapphire: It was American, and it was guaranteed completely natural.

The Yogo sapphire had been through a rough journey in the gem world: it had been slighted at various times for being "too small" or "too flat"; its origin had been misrepresented; that earthy frontier name of Yogo could never hold a candle to the exciting and exotic names of Siam and Ceylon; finally, the long line of American failures at the Yogo dike implied to many that perhaps Yogo sapphires simply weren't competitive with other stones. Now, the old Yogo was coming back as the Royal American Sapphire. The image of one of the world's greatest precious gems was about to be redeemed in grand style.

Charles Gadsden would have loved the next act.

Chapter 6

Rocking the Boat

A quarter-million gem cutters make Bangkok, Thailand, the center of the world's colored gem industry. Ninety percent of the world's mine production of rubies and sapphire, whether from Thailand, Ceylon or Australia, is cut in Bangkok, then sold by Thai dealers to gem merchants around the world. When the turbulent economy of the late 1970s drove prices of gold and precious gems to record levels, Thai gem dealers were understandably ecstatic. But their smiles soon faded when the booming demand for fine blue sapphires outpaced the mine supplies.

Sapphires were being mined in huge quantities, especially in Sri Lanka (Ceylon) and Australia, but few were of the classic blue colors that commanded the top prices. The majority of Australian sapphires were exceedingly dark and tinged with disturbing heavy tones of greens and yellows. The sapphires of Sri Lanka occurred in a very broad range of colors, relatively few of which were desirable blues. A large percentage of the Sri Lankan sapphires were a milky white. Known locally as *geuda* (gay-oo-dah), Sri Lankan sapphires had no market and very little value as gemstones. *Geuda* was considered "junk" and never entered the gemstone market. In testimony to its low value, *geuda* was often stored in buckets and old fuel drums. Wealthy Sri Lankan Muslims did manage to find an interesting use for *geuda*; some purchased the stones by the bucket, using them to landscape their ceremonial gardens.

It was no secret among gemologists and gem merchants that the color of certain natural gemstones, including the corundum gems, could be altered or enhanced by heating. Marco Polo wrote of the heating of a pink sapphire in Ceylon in 1271. So, too, did Jean Baptiste Tavernier, a well-traveled seventeenth century French gem merchant. An 1860 British government report on Ratnapura, the primary Ceylonese gem mining area, tells of ruby heat treatment.

> The blue tinge which detracts from the value of the pure ruby, (whose colour should resemble "pigeon's blood") is removed by the Singhalese by enveloping the stone in the lime of a calcined shell and exposing it to high heat.

By 1900, mineralogists were heating gemstones to study the phenomenon of color alteration from the scientific standpoint.

More recently, heat treatment has become of considerable importance in preparation of gems for marketing. Aquamarine, a gem variety of the mineral beryl, in its natural state almost always has a strong greenish hue. Virtually all the aquamarine sold in the last forty years, however, has been heated to remove the yellow color component, thus producing stones with the classic, clear, and highly desirable "aquamarine" blue which the buyer has come to expect.

While heat treatment can enhance the color of such stones as aquamarine, it can actually "create" entirely new colors in others. The best example would be zoisite, a calcium aluminum silicate, mined in Tanzania, Africa. Natural zoisite occurs in very unappealing colors ranging from yellow-green to purple-brown. Heat treatment, however, will conveniently change the ugly zoisite colors to deep, lovely blues, colors some jewelers have compared to those of Kashmir sapphires. The reborn zoisite was quickly given a more appealing name—tanzanite—and successfully marketed by Tiffany & Co. at prices commensurate with its new chic appearance.

Since heat treatment affected not only the color, but also the *value*, of gemstones, the Federal Trade Commission, in the interests of consumer protection, addressed the issue of whether artificial gemstone color alteration had to be disclosed to the consumer. In its 1957 Jewelry Industry Trade Practice Guides, the FTC declared that only gem treatments that were impermanent or which could be detected by a gemological laboratory required consumer disclosure. Heat treating of aquamarine, already a gem industry "tradition," was considered permanent and undetectable. Later, when tanzanite emerged from the ovens, it, too, was judged to meet the standards for nondisclosure. Aquamarine and tanzanite quietly became precedents for a non-official, general policy of nondisclosure. Within the industry, a broad range of treatments and treated gems became "acceptable" and "traditional"—of which the consumer was told virtually nothing.

That necessity is the mother of invention was proved again in the late 1970s in Bangkok. Instead of merely lamenting the shortage of classic blue sapphires, the Thais did something about it. If blue sapphire cannot be mined, the Thai gem merchants thought, perhaps it can be made. Accordingly, they began experimenting with heat treatment of poorly colored sapphires. There are many amusing, fanciful tales of just how they achieved success. Since none are documented, only one fact remains: a nameless Thai gem dealer managed to transform a worthless, off-color corundum crystal into a marketable blue sapphire. What followed was inevitable.

The Thais knew the greatest source of off-color, dirt cheap sapphires in the world were the heaps of Sri Lankan *geuda*. Descending on Sri Lanka in hordes, they bought up every bucket of *geuda* they could get their hands on. Back in Bangkok, the *geuda* was heated, reheated and, if necessary, heated again. Much was ruined, but the rest emerged as blue sapphire ready for the cutters, then for the markets. At first, the Sri Lankans had been delighted to unload their worthless *geuda* on the Thais. But when they figured out what was happening, their prices went up. By 1980, *geuda* cost nearly as much as rough blue sapphire. By that time, however, the Thais had staged one of the greatest "killings" in gem trade history. They had bought millions of carats of junk at bargain basement prices and, with a little heating, turned much of it into high-priced blue gem sapphire.

The early Thai heating techniques and equipment were primitive, but served their purpose. The most popular oven was an old fifty-five-gallon drum with a ceramic lining and a door at the bottom. A crucible containing the sapphire roughs was placed in a bed of glowing embers within the drum. Fuel was a matter of convenience and supply and included everything from coal to charcoal, wood, coconut shells and, occasionally, dried water buffalo dung. Time and temperature, as well as the critical time of cooling, were left to the discretion of the particular oven operator. Some Thais packed the roughs in various materials to protect them and, hopefully, improve the results through altering the heating environment. Thai heat treating was much more of an "art" than a science; any combination of time and temperature that worked was guarded zealously. Some of the *geuda* melted, exploded, shattered or was otherwise damaged beyond salvation; in other cases the resultant color was worse than the original. But when the process worked, color and color intensity could be improved as much as 80 percent. Given the inexact, makeshift methods, a specialized vocabulary evolved to describe the Thai heat treating processes. Sapphires passing through the Thai "kitchens" were said to be "cooked," "fried" or "burned."

In heat treating, nothing was added and nothing was synthesized. The raw material, *geuda*, chemically and physically, was truly sapphire; it was an aluminum oxide crystal containing the necessary chromophores—iron and titanium—to *potentially* make a blue sapphire. But nature, in her infinite wisdom, had chosen not to provide the necessary heat and pressure to properly arrange the chromophores within the crystal lattice to create a blue color. Heating the *geuda* to temperatures around 1800 degrees Celsius—nearly the melting point of corundum—altered the covalent bond of the iron and titanium oxides and better distributed them within the lattice. Theoretically, heat treating was duplicating the process nature would have used if she had decided to.

Inspired by their profits with heated *geuda*, the Thais directed their inventiveness to other sapphires. They found they could heat treat the dark Australian stones, reducing the color intensity by about 10 percent and greatly improving marketability. Heat treating became the norm. Thais were heating virtually all sapphires whether they needed it or not on the pretense of "making something good a little better."

Incredibly, by 1980, *about 95 percent* of all the sapphires on the market had been heated, both those on the retail counters and those in the enormous gem inventories of American gem merchants. Yet the issue of disclosure had been ignored. Gem trade spokespersons told their troubled peers—not the public—that, yes, almost all their sapphires were heated, but it was "acceptable" and "traditional."

Furthermore, disclosure wasn't necessary, it was not illegal, or a violation of the public trust, for heat treating of sapphires was permanent and non-detectable. At any rate, since the industry was sitting on a king's ransom in heat treated sapphires, this was certainly not the time for a public airing of the issue.

But disconcerting whispers were already being heard around the trade. In 1981, the American Gemological Laboratories, of New York City, announced that they could detect heat treatment in sapphires 95 percent of the time. The gem trade immediately went on the defensive, noting that 95 percent was not *all* the time. Another disturbing question was that of permanency. Most gem dealers and jewelers claimed, assumed or prayed that the color of heat treated sapphires was most certainly permanent. No one really knew if the heat treated *geuda* was permanent or not over a period, say, of ten or fifteen years. It was simply too soon to tell. But there were reports from respected gem dealers who had seen heat treated sapphires revert back to their original color.

Since there was no industry standard for disclosure, that question was left up to the poor retail jeweler. Since 95 percent of all sapphires were cooked, that meant 5 percent were not cooked. And since only a gemologist, not a jeweler, could distingush between the two, what was there to disclose?

To complicate the heated sapphire issue was the question of value. Along with beauty and durability, rarity is one of the factors that give a precious gem its value. Until the 1970s, the market value of sapphire was determined essentially by supply and demand. The classic blues, quite rare in nature, naturally were the most valuable. In the past, "traditional" heat treating was employed to slightly enhance the color and beauty of already valuable gemstones, such as removing the small greenish tinge of a fine ruby. But the Thai mass heating of *geuda* that began in the late 1970s was a marked departure from "tradition." Already valuable gems were not simply being enhanced; they were being *created* out of worthless junk—and sold to consumers at the same prices as fine, natural gems. Their value had not been created by nature as a function of rarity; it had been artificially created in a Thai oven. The element of rarity had plainly been tinkered with. Selling cooked *geuda* to the retail consumer for the same price as natural sapphires had major ethical ramifications.

By 1982, the gem trade had gotten itself into an awkward, embarrassing and potentially disastrous position. It remained defensive, carefully limiting the growing heat treatment debate to trade circles. It was only too aware that an adverse mass reaction among consumers could be catastrophic, both to the value of enormous inventories of cooked sapphire and to industry trust and credibility. The trade found the treatment issue to be quite serious, but nothing that might not be worked out in time. Perhaps in ten years, the issue would resolve itself; by that time, even mass heating of *geuda* might be "traditional." Everything would be fine as long as nothing came along to rock the boat.

The boat started rocking when Intergem began test marketing the Royal American Sapphire line in 1982. The Royal American Sapphire, of course, was the same Yogo sapphire that had been around for eighty years. But now, the Yogo stood apart, clearly different from every other sapphire in the world. It would be unthinkable to artificially alter the magnificent cornflower blue color of a Yogo sapphire. In Bangkok, a Thai oven operator would be considered a master if his cooked stones even approached the blues of the Yogos. As the Yogos hit the market, complete with the printed guarantee that they were completely natural and untreated, time began running out for the gem trade to resolve the treatment issue. Heat treatment began receiving regular coverage in the trade journals. There was no avoiding the directness of comments such as those of New York gem merchant Reginald C. Miller that appeared in a 1983 issue of *Goldsmith*.

> ... During this last couple of years we have had a very big problem with some colored stones caused by treatment. I find it very difficult to sell a stone that has been burned, especially when asked to have a stone certificated for verification on this point. In fact, many of us have decided not to get involved anymore in possibly burned stones.
>
> I myself have been burned almost as many times as the stones have.
>
> As knowledgeable as I have considered myself to be, I have bought stones that have been treated in such a manner, that upon recutting or exposure to everyday light, they have lost their color and became almost worthless—especially when compared to their color and brilliance when I first purchased them. This has made not only me, but many other dealers absolutely wary of being further involved in some colored stones.
>
> I have spent years educating the gem trade about the beauty and rarity of yellow sapphire, only to be made a complete idiot when they

were returned because they had faded.

Everybody feels like a fool. We feel as though we have cheated the person we have sold to. In some instances, the people who have sold the stones to me have made good, but there were some people who refuse to admit that they have sold a stone that has been tampered with. . . .

The same thing applies to blue sapphires. I can remember buying stones that I thought were Cambodian that were a strong, but good, blue.

The remainder of some of those lots I have today, and for the life of me I could not imagine why I would buy such black stones. They were not black when I bought them, they were blue. In the meantime, they reverted to a dark color.

I have since come to realize that these stones were not Cambodian, but typical Australian sapphires.

Most of us are well acquainted with that blackish, greenish-blue of the Australian sapphire. There is nothing wrong with them; they just happen to be a color that is hard to sell in the United States.

No wonder we are wary of treated stones!

In May, 1983, *Jewelers' Circular/Keystone* ran a very lengthy and comprehensive article titled "Treated Gems: Why Not Face The Facts?" The article opened with a revealing statement about the extent of treated gems on today's market.

If a white tornado whirled through your store and removed all of the treated stones, you'd probably find only about one colored stone in four left to sell.

Surprised?

It's a fact of life. World supplies of fine gemstones are dwindling. Treatment is being used to fill the gap.

If not for treatment, dealers would be selling far fewer colored stones. There just wouldn't be enough beautiful, saleable gems to go around. . . .

The article then mentioned another of those recurring stories about the reversion of a treated sapphire.

A retail jeweler bought three small yellow sapphires from a reputable dealer. He mounted the nicest one—an intense canary yellow—in a ring with a couple of diamonds. The piece carried a retail tag of $700.

The man who bought it told the jeweler he had never gotten his wife anything really nice before. The ring was a special surprise for her birthday.

A few weeks later, the man brought the ring back. The sapphire had faded so badly it was almost colorless. His wife was shocked and disheartened.

Isolated incidents?

Hardly.

Many jewelers and gem dealers defended gem treatment, arguing that anything done to enhance beauty was to the consumer's advantage. They also cited tradition and the fact that heat treatment was itself a "natural" process, as in these examples:

> The change in heated sapphires, for example, comes from the capacities within the gem. Heat is needed to finish what Mother Nature didn't. She stopped short of her final work, so to speak. Man has found a way to bring out what might have come out if the stone had been given the chance a million or two years ago when it was created.

> Heat treating is nothing new, it's been done for centuries. It's a tradition and should be accepted as one.

> What we're really doing is selling beauty. If heating a stone makes it more beautiful, what difference does it make?

> If the color is permanent and the treatment can't be detected, what difference is there from an untreated stone? What is there to disclose?

Treatment did make a difference to dealers, especially in the matter of value. *Jewelers' Circular/Keystone*, in its published treatment debate, pointed out two instances.

> **A PORTRAIT OF AMBIVALENCE**
>
> "Suppose you have two sapphires," explains Abe Nassi, A. Nassi, Inc., New York. "One is not treated and priced at $5,000. The other is simply heated. It has the same color, quality and permanence, and looks the same in all light as the other stone. I see no reason why it should cost less."
>
> "Man is taking the same process as nature," Nassi adds. "He's putting the stone into an oven, so instead of taking 100 years, it's taking a few hours. I'm pushing the temperature almost to the melting point. I'm just helping nature."
>
> The change in the stone is not such a major thing, he continues. You must often go through complex gemological tests to determine it. Even then, he says, it is not always detectable. If the color will never change, what difference will it make?

Apparently it made a good deal of difference to Nassi himself, for the interview concluded with this interesting statement.

> Yet Nassi himself admits that when he buys, he won't pay the same for a treated stone.

On the same page appeared another statement implying the existence of a two tier pricing system within the trade.

IGNORANCE CAN COST YOU

> Jack Abraham (a New York gem dealer) tells how another dealer offered him an eight-carat sapphire. It had the glorious color you'd expect from a Kashmir, he says.
>
> "The price was right. The stone was nice." It even had a certificate stating it to be a Kashmir. But something was wrong.
>
> "It just didn't look right to me." Abraham recalls.
>
> So he took it to a third dealer, one who handles a lot of Kashmir material.
>
> "What do you think?" Abraham asked him.
>
> "The price is okay," he responded.
>
> Still concerned, Abraham took the stone to the Gemological Institute of America's Robert Crowningshield and the American Gemological Laboratory's Cap Beesley.
>
> They both told him the stone had been heated. It wasn't a Kashmir at all. It was a Ceylon stone that had met with particular success in the fire.
>
> "The point is," Abraham explains, "a lot of people would have bought that stone for a Kashmir. A dealer would have paid the $8,000 or $10,000 a Kashmir like that would have been worth." As a Ceylon stone, Abraham put its value at half.

Gemologists, jewelers and gem dealers all offered their opinions on the gem treatment debate. Another group that also stood to gain or lose in the matter was not asked for its opinion—the consumers who supported the multi-billion dollar jewelry industry. To the average consumer, the purchase of a precious gem marked a significant event in his or her life. The stone itself, for which they were paying their trusted jeweler a substantial amount of money, was automatically assumed to be a rare, durable and beautiful creation of nature, not something that had been toyed with by man for purely commercial purposes. Among the first to speak out for the retail purchasers of precious gems was Rose Leiman Goldemberg in her book *All About Jewelry* (Priam Books, 1983). Speaking not as a gem dealer, but as a gem consumer, Ms. Goldemberg's views probably reflected those of millions of other gem consumers.

> In general, synthetic stones may look very nice, but are not at all the same as natural ones, and any jeweler who tries to beguile you with nonsense like, "They have the same chemical structure as the real stones; what's the difference?" should certainly not be trusted. The value of a synthetic stone is much less than that of a real one; but perhaps even more important, part of the mystique and wonder of a natural stone is its individuality, its beauty, and all of the care and craft and passion—of man and nature—that brought it out of the earth, out of the hands of craftsmen from foreign lands, to rest, blazing and safe on your finger. To claim that a synthetic stone is as precious as a real one is a little like saying a TV dinner is as good as a home cooked meal.
>
> Another way of faking gemstones—a method jewelers seldom discuss—is to apply heat to stones of poor color to develop a look that is more marketable. Most sapphires, when they are mined, are an unappetizing milky gray; the blue stone, or even the clear-colored stone of any hue, is a rarity. These ugly stones, called geuda, are sometimes subjected to experiments in which heat is applied to improve color and clarity. The dealers in Hong Kong and Thailand "cook" geudas to make them more marketable, heating them up to about 1,700 degrees centigrade. This technique either ruins them or results in stones of better hue, usually yellow or blue. Now these heat-treated sapphires are genuine natural sapphire; all the tests will prove that. So how are you to know if they have been treated? And does it really matter?
>
> Well, it matters to me, because part of my pleasure in a gemstone is the astonishing set of natural coincidences that contributed to its creation. But I can see the other side of the argument. The digging and cutting and polishing and selling of any stone are certainly not "natural" or coincidental. No gemstone ever leaped, fully faceted, into its owner's palm! But, nevertheless, a treated stone is generally considered in this business to be faked and the price should reflect that fact. . . .
>
> . . . The issue of cooking a stone can perhaps be compared to getting a face-lift. Is it wrong to deceive, or is it a legitimate cosmetic procedure, solely designed to enhance beauty and pleasure? Well, I would rather have a totally natural stone, and I always ask the question directly: "Has this stone been heat-treated or dyed?" and make sure a GIA (Gemological Institute of America) certificate accompanies each major gemstone purchase. But the sad truth is that many dealers don't know, or don't always want to know, the history of a stone—and many don't

consider it necessary to tell you even if they do.

As the gem trade reeled with debate, uncertainty and even outright fear, Intergem was slowly expanding the retail sales network for their Royal American Sapphire. For the first time, both retail jewelers and consumers had an alternative to the usual stocks of heat treated sapphires. Some jewelers saw certain advantages in the Yogos, especially those who felt uncomfortable with the question of disclosure, or who feared heat treated sapphires might later come back to haunt them. To many "mom and pop" jewelry stores, where an image of community trust had been built over twenty or thirty years, the Yogo sapphire represented honesty, confidence and peace of mind. By summer, 1983, over 200 retail jewelers were carrying the Royal American Sapphire line.

With the reappearance of the guaranteed untreated Yogos, the issue of heat treatment was slowly going public. This article was published in *Southern Jeweler* in August, 1983.

HEAT-TREATED GEMS A HOT NEW ISSUE

According to Dennis Brown, president and chief executive officer of Intergem, Inc., of Denver, producer supplier of 100% natural sapphires, the issue of heat-treated gemstones has turned into a hot debate. "Even those in the gem industry who have chosen to ignore the controversy stand to get burned," he warns.

Basically, two factors must be evaluated in the controversy: value and the consumer's right to know. "Value is an extremely important issue that should be settled within the industry before it becomes a public debate," Mr. Brown says. Heat treatment has been used to manipulate the color of gemstones in certain southeast Asian countries for more than a century. However, in the last decade, increased consumer demand has resulted in indiscriminate use of heat treatment to improve color of otherwise worthless stones; these stones—called geuda—then are marketed at a price often competitive with natural gems. A serious consequence is that the market is flooded with artificially colored stones. Rarity, which once gave precious stones their value, is no longer a factor when distinction is not made between treated and natural.

Concerned with the consumer's right to know, Intergem issues a certificate of guarantee that establishes the authenticity of origin of its Royal American Sapphires featured in the jewelry line the company produces and distributes. The guarantee is an assurance of the untreated beauty of the gemstone, since the company believes the consumer would rather prefer a natural gemstone to a treated one of the same price. Because a treated sapphire could revert to its unattractive original color, "it doesn't take an expert to realize which gemstone is more valuable," Mr. Brown says.

However, jewelers and dealers have been reluctant to offer information on heat treatment for fear that it might discourage sales. "Nevertheless, consider the repercussions when consumers find out that their precious stones have been treated," Mr. Brown points out. "Now that detection is 95% sure through laboratory tests, the industry could experience a consumer backlash."

Intergem's stated concern for the interest of the consumer may have been a bit idealistic; doubtlessly, the primary concern was expanding the market and eventually achieving a profit, like every other gem dealer in the business. But the company was on a collision course with the dominating interests of the gem trade and "on the side" of the consumer. The "untreated" quality of the Yogo sapphire was becoming a big marketing advantage. What was best for Intergem was not best for the gem trade, however. Intergem's official position on heat treatment was that it was fine—as long as it was disclosed. Full disclosure, of course, might mean a two tier pricing system, with natural sapphires on top. Fortunately for Intergem, the only sapphire that could be certified *en masse* as untreated was the Yogo. While Yogos were becoming increasingly controversial, they were also becoming more marketable. By the end of 1983, Intergem's network of independent dealers exceeded 400 stores. Sales that year totaled 7,000 pieces of jewelry containing over 2,000 carats of Yogo sapphires. The wholesale value was $1.6 million—an eight-fold increase over the year of test-marketing. On August 1, 1984, the *National Jeweler* reported that Intergem was indeed "forcing" the issue of heat treatment, not only on the industry, but on the public.

INTERGEM RAISES TREATMENT DISCLOSURE ISSUE

NEW YORK—Intergem, a Colorado-based firm that sells untreated gemstone jewelry, is forcing the industry to face the complicated issue of gem-treatment disclosure—perhaps more quickly and more publicly than it would like.

As part of its marketing campaign, the firm has invited representa-

> tives of the consumer media, including the prestigious *Wall Street Journal* to visit its Montana mine and write about its operations. Because the linchpin of Intergem's marketing program is its guarantee that its stones are untreated, these meetings with consumer media make it likely that the disclosure issue will soon "break" in the consumer press.
>
> This guarantee, which appears in Intergem's trade advertising, is controversial because gem dealers have long maintained that treatment cannot always be detected. They have used this fact to justify nondisclosure.
>
> The American Gem Trade Association, which represents more than 400 gem dealers, including Intergem, took issue with the firm's advertisement. In a letter dated May 31, Nace Moghadam, chairman of AGTA's grievance committee, stated, "I would hope you withdraw this advertisement and reword it so that the integrity of your fellow dealers is not compromised."
>
> As a result of ensuing discussions between Intergem executives and AGTA head Roland Naftule, a meeting has been planned to discuss gem treatments. Participants will include AGTA leaders, Intergem head Dennis Brown and Gemological Institute of America officials. . . .
>
> Brown said that both Intergem and AGTA would like to see the industry address the issue of treatment disclosure. He noted however, that the New York meeting and, particularly, AGTA's invitation for Brown to attend, may be partly attributable to the industry's realization that the public is going to learn about treatments "sooner rather than later."
>
> Naftule denies that the colored-stone industry has anything to fear from an open discussion of gem treatments. He noted that treatments have been widely discussed at industry shows and also at recent gemstone conferences in Acapulco and Tel Aviv.

Gem treatments had indeed been "widely discussed," but never with the consumer. But that was about to change. The *Wall Street Journal*, sufficiently impressed both with Intergem's steady growth and its considerable impact on the gem trade, sent a reporter to Yogo to get the story firsthand. Seventy years earlier, Charles Gadsden tried to drum up a little publicity for Yogo sapphires by placing handout literature promoting the "Montana Blue" on Great Northern Railroad passenger trains. The spirit of that somber-faced Englishman, who some locals say still watches over the Yogo dike, must have smiled on the morning of August 29, 1984, when Yogo sapphires hit the front page of the 2.5 million circulation *Wall Street Journal*, the nation's biggest and most influential business daily.

CARATS AND SCHTICKS: SAPPHIRE MARKETER UPSETS THE GEM INDUSTRY

NATURAL YOGOS CLAIM IMPLIES HEAT-TREATING IS WRONG: BUT PRACTICE IS DEFENDED

YOGO GULCH, Mont.—. . . Suddenly, however, the gem industry is paying plenty of attention to the goings-on at Yogo Gulch. Intergem, Inc., a Denver-based company that bought the mine—actually a five-mile-long crack in the earth's surface—four years ago, has big plans for the Yogo. The cornflower-blue sapphires have been introduced into more than 600 jewelry outlets in the past year, including trial runs with such mass-merchandising giants as Zale Corp., May Co. and Montgomery Ward & Co. [reporter's error: the third was J. C. Penney & Co.]. Next year, Intergem plans to try out a national advertising campaign to introduce Yogos into the wedding- and engagement-ring markets.

That could mean a head-to-head marketing battle with giant De Beers Consolidated Mining Co., whose "diamonds are forever" ads are the only other all-out marketing effort in the gem business. But Intergem isn't waiting for that confrontation to shake up the $8.6 billion-a-year U.S. jewelry industry.

The company's current marketing touts its Yogo sapphire—now renamed the Royal American Sapphire—as "natural" and offers customers written guarantees that the sapphires haven't been treated with heat or chemicals. The campaign has thereby aired a fairly well-kept secret: Though jewelers rarely tell their customers, almost all the colored gems these days including top-grade rubies, sapphires and emeralds, undergo "color enhancement" by heat or chemicals before they are imported into the U.S. . . .

The practice has triggered a sharp debate within the industry. "The consumer should get what he thinks he's getting," says Joel Windman, an attorney who heads the Jeweler's Vigilance Committee, an industry ethics group that has proposed full-disclosure guidelines to the Federal Trade Commission. "When you pay for a Cadillac," Mr. Windman says, "you don't want to get a Chevrolet in disguise."

Intergem officials contend they are simply following a policy of consumer honesty in their sapphire promotion. "We want to nudge the industry to come to grips with the issue of non-disclosure," says Steven M. Droullard, Intergem's marketing vice president.

Whatever the motive, the strategy seems to be working. Intergem sold $1.6 million of Yogos in 1983. It has already sold $1.4 million of them this year, with heavy pre-Christmas sales likely to increase that figure by 70% by year end, according to company officials.

Some gem dealers contend that Intergem is making mountains out of molehills with its Yogo promotion. "Heat treatment is completing what nature didn't get a chance to finish," says Roland Naftule, a Phoenix, Ariz., gem wholesaler. Mr. Naftule is the president of the American Gem Trade Association, a trade group that includes about 400 sapphire wholesalers.

He says that the practice of touching up certain precious gems goes far back. Centuries ago the gem trade discovered that pearls fed to goats would eventually be made brighter and cleaner than before. "It's fabulous what man and nature can do working together," Mr. Naftule says.

Even Tiffany acknowledges that it sometimes sells its high-rolling custom-

> ers heat-treated stones. "It's absolutely common practice," says John Loring, Tiffany's design director and senior vice president. "We've published enough articles on the subject that the public should be aware that certain stones are heat-treated." Mr. Loring says that Tiffany tells its customers about the processes if they ask.
>
> Yet the subject is a touchy one. Few jewelers volunteer the information to customers; some say they don't always know themselves. And many dealers aren't pleased that a newcomer like Intergem has decided to raise the issue. . . .

While treatment was the hot new issue among gem dealers across the country, some things at Yogo seemed timeless. When *Wall Street Journal* reporter Bill Richards visited Yogo, he was invited to pick through the jig concentrates to test his luck as a sapphire miner, just as other visitors had done decades earlier at the English Mine. In those days, of course, it was Charles Gadsden who relieved the visitors of their finds. This time it was Harry Bullock, in an incident Richards included in his article.

> We poke through several large baskets filled with gravel as a jet of water washes away the clay and heavier rocks. The gravel is studded with hundreds of small blue stones. With tweezers, I pick out one the size of my thumbnail. Mr. Bullock, a former hard-rock mining engineer, estimates the stone will cut to 1½ carats, worth perhaps $3,000 when set and sold retail. The blue stone is gently relieved from my grasp and dropped into a sack the size of a bowling bag, filled with uncut gemstones.

The *Wall Street Journal* article, like the treatment issue itself, was the subject of some controversy. John Loring, of Tiffany & Co., claimed he was "misquoted and misrepresented totally," that his discussion had concerned only aquamarine and tanzanite, and that the use of his quotes in a sapphire article was "inappropriate." The *Wall Street Journal* reporter, however, stood by his story.

The Yogo sapphire had indeed rocked the gem trade's boat and Intergem had made few friends among the dealers. For two years, Intergem stood alone as the only company formally fighting for trade recognition of a "difference" between treated and untreated sapphires. Was it only a marketing ploy to gain publicity for the Yogos, or did a difference really exist?

The answer became apparent in the January, 1985, *National Jeweler*. Reginald C. Miller, the prominent and respected New York gem dealer who once said he had "been burned almost as many times as the stones have," ran a full page ad directed at retail jewelers.

> **REG MILLER BELIEVES THERE IS A DIFFERENCE BETWEEN TREATED AND UNTREATED SAPPHIRES! DO YOU?**
>
> Fine jewelers build their reputation on the quality and the reliability of their stones, not the low prices. And, only untreated natural color sapphires can truly be considered "fine."
>
> . . . Please call and let us explain the availability and varieties of untreated sapphires. We can show you a world of difference.

If, as Reginald Miller said, there was a "world of difference" between treated and untreated sapphires, Intergem would be in a stronger position than ever, for the Yogo dike was the biggest and most reliable source of untreated sapphires in the world.

Chapter 7

The Great American Sapphire

By 1984, ninety years after Jake Hoover panned those blue pebbles, the Yogo dike, under the ownership and operation of an American company, had achieved recognition as one of the world's most important precious gemstone sources. And the Yogo sapphire was emerging as a standard of quality among all sapphires. The return of the Yogo was made possible by a combination of business acumen and a fair amount of luck. Intergem had begun with a quality product and backed it with expertise and excellent management that started at the mine and ended at the retail jewelery counter. The luck was a matter of timing; Intergem had benefited from the rise in value of precious gems, the growing popularity of colored gems, a new market demand for smaller gems, and, finally, the great amount of publicity generated by the debate that still raged over the controversial practice of treating gems.

During the 1930s, few people drove down the dusty road from Utica to visit the Yogo dike. Even that had changed for, as Bill Richards had written in the *Wall Street Journal*, people were now "paying plenty of attention to the goings-on at Yogo Gulch." The Yogo sapphire had already appeared in over forty articles in gem trade publications and was receiving increasing attention from the consumer media. Visitors now included staff reporters from the *Denver Post's Empire*

Magazine and camera and production crews from NBC's *PM Magazine*.

The largest group of visitors to the Yogo dike was made up of retail jewelers and other gem trade professionals who responded to this Intergem invitation:

> **MINE TOURS!**
>
> This summer a limited number of jewelers will be able to visit the fabulous American Sapphire Mine at Yogo Gulch, Montana. They will see Yogo Creek where Jake Hoover first discovered sapphires a hundred years ago and stand on the ridge where Charles Gadsden sorted sapphires for the London-based New Mine Sapphire Syndicate in the early 1900s.
>
> Pick up your own sample of blue ground ore and experience first hand the world's largest and most important sapphire deposit. . . .

Over 500 gem trade professionals accepted that invitation in the summer of 1984. For many, it was their only opportunity to ever visit a commercial precious gemstone mine. Intergem made the best of it, effectively using the tours to stimulate interest in the Royal American Sapphire line and to expand its network of independent authorized retail dealers.

In 1983, Intergem had acted to raise additional operating capital and to provide the original limited partners with a greater degree of investment flexibility. Newport Oil and Gas, Inc., a small Utah-based energy firm, was acquired as the simplest means of going public. The new company, Intergem, Inc., began over-the-counter stock trading as IGEM. Even with the steadily increasing business growth and stability, as well as continuing publicity, not everyone in the gem trade was willing to accept or befriend Intergem and its Royal American Sapphire. Typical of the criticisms leveled against Intergem and its product were these comments that appeared in the Christmas, 1984, newsletter of the House of Onyx, a leading importer of gemstones and decorative stones.

> There has been quite a bit of publicity lately about what is called the YOGO sapphire, and most of it is quite confusing.
>
> The YOGO gulch, in Montana, has produced some small sapphires for about a hundred years. Many firms have been involved in this over the past century, most of which never made any money and most of which lost everything they had. That history should tell you some-

> thing. The current operator, Intergem, Inc., of Denver, has been pushing their "American" sapphires in a lot of new ads. Most of the ads have a point in telling the public that most colored stones are heat treated, or treated in some way. This has not gone over well with most dealers as the ads lead the public to believe there is something wrong, and/or sinister, with heated, and/or irradiated gemstones.
>
> . . . Permit me to say that all legitimate stone dealers have always known of the Yogo sapphires—and all international dealers have considered them to be too expensive. The higher price is due to American costs of mining the stones, which is not competitive. This is the main reason there are not many stones, of any kind, cut and mined in the USA. There are many sapphires from Ceylon, of the same color and brilliance at cheaper prices. The Ceylon sapphires also come in much larger sizes.
>
> They (Intergem) claim sales so far this year (the first half of 1984) of $1,500,000.00—hardly much of a figure in the stone business.
>
> . . . According to these figures quoted, I personally know over a dozen stone dealers, here in the States, that carry more stones with them in their briefcase, than Intergem claims as their total net worth. Let's all wait a year or two to see what the company looks like at that time. Remember, good, well priced stones sell themselves and it doesn't matter where they are from—Montana, Ceylon, Africa or Burma.

Harry Bullock, now chairman of Intergem, Inc., accepted such criticism by paraphrasing P. T. Barnum. "There is no such thing as bad publicity," Bullock stated, "as long as they spell your name right." The publicity, both good and bad, certainly seemed to be doing its job. By December, 1984, Intergem's sales reached $3 million, double that of 1983, and 1,000 stores were now carrying the Royal American Sapphire line. Chain outlets were now playing a bigger role in retailing and included such prestigious names as Saks Fifth Avenue, H. Stern, Macy's and Zales.

During its first four years, Intergem had demonstrated rapid growth but still could not be considered a major factor in the industry, as gross annual sales amounted to less than one percent of the overall national sapphire jewelry market. Bullock confidently felt that the admittedly small part of the market pie left "just that much more room for future growth." Intergem and the Yogo sapphire had earned the front page of *The Wall Street Journal* not because of size, but because of impact on the gem trade and, most importantly, potential for the future.

Buoyed by strong Christmas sales, Intergem began the new year on a very optimistic note, and for good reason. Projections indicated another year of sharp market expansion to a new annual sales level of 25,000 pieces of jewelry containing 8,000 carats of cut Yogo sapphires with a wholesale value of well over $6 million. President Dennis Brown was even offering long-term market projections, believing a realistic market target for the year 1990 to be fifteen percent of the nation's 35,000 jewelry stores, or over 5,000 retail outlets that would be carrying the Royal American Sapphire line.

Much of Intergem's future rested on its most important asset—the Yogo dike, a mine unique among the world's sapphire sources. Unlike foreign sapphire deposits, both historic production and reserves were known. The total production of the English Mine and all subsequent American mining efforts through 1980 was about 20 million carats, including both gem and industrial sapphire. Intergem, in its first five years, had mined only about 750,000 carats, some 40 percent of which were gem quality. Yet the total historic mining effort had literally only scratched the surface. Intergem's stated sapphire reserves—based only on a depth of 500 feet—totaled about 40 million carats of sapphire with a market value of well over one billion dollars. The actual depth of the dike, however, was at least 7,000 feet. Without exaggeration, the quantity of gem-quality blue sapphires contained in Montana's Yogo dike exceeded the sum total of those remaining in all of the world's other known sapphire deposits.

Intergem was already planning for the day in the not-distant future when surface mining at the dike would no longer be practical or economical. The transition to underground mining called for a unique mining method designed specifically for the configuration of the dike and the nature of the dike rock. Two shafts would be sunk into the dike and conventional mining methods used to prepare preliminary workings. Overhead supports set into the adjoining limestone walls would suspend a moveable carriage. From the carriage, miners would direct high pressure water jets across the "face," or vertical working surface of the dike rock. The 600-pound-per-square-inch water jets would break down the *in situ* dike rock into a slurry which would fall to the bottom of the large underground "room," a working unit up to 200 feet long. As the slurry fell to the lowest levels, grizzlies—heavy steel screens—would retain the large fragments. The slurry, containing the loose sapphires, would then pass through a series of gravitational and sizing jigs where the stones would be recovered. The many advantages to this proposed system of underground hydraulic mining included minimal surface environmental disruption, elimination of mechanical mining and hauling steps that might damage or destroy the sapphires, and the inherent establishment of a perfect security system, for sapphire recovery would take place in a single, easily controlled area of the underground workings. A final advantage would be elimination of surface disposal problems, for the tailings would become "backfill," a slurry used to fill in mined-out areas of the underground.

While Intergem planned mining operations at the Yogo dike and further market expansion from its Aurora, Colorado headquarters, it received an unexpected, but welcomed, wave of publicity that began in Philadelphia, Pennsylvania. Television reporter and consumer advocate Herb Denenberg, intrigued by *The Wall Street Journal* article on Yogo sapphires and heat treatment, had launched his own investigation into what he felt were fraudulent and deceptive practices in the jewelry trade. Among Denenberg's major points was the matter of gem treatment nondisclosure.

Using wired "customers" and hidden video cameras, Denenberg recorded a dramatic exposé. When the "customers" asked specifically if a certain gem had been treated, jewelers often brushed the question aside with comments like, "There's no need to worry about that" and, "Oh, that's not important at all." In some cases, jewelers flatly stated their gems were not treated when, in fact, they had no way whatever of knowing it. Denenberg's exposé, which was a huge embarrassment to the jewelry industry, aired in February, 1985, on Philadelphia's WCAU-TV, a CBS affiliate. For his featured interview guest, Denenberg chose none other than Steven Droullard, Intergem's vice president for marketing. Droullard, of course, was eager to present Intergem's position that gem treatment was fine, as long as it was accompanied by full disclosure in the interest of consumer protection.

Many television viewers, especially those who had no idea of the extent of artificial, undisclosed gem treatment, were shocked, among them Pennsylvania State Senator Stewart J. Greenleaf. Two months later, Senator Greenleaf shocked the jewelry industry, which had yet to recover from the Denenberg television exposé, by introducing legislation calling for mandatory disclosure of all gemstone treatment to consumers. The bill, Pennsylvania State Senate Bill Number

827, provided for civil action "to recover the price paid for the gemstone, any costs incurred in determining that the gemstone had been treated and further damages in the amount of 10 percent of the purchase price or $200, whichever is greater."

In response to the new legislation, Steve Droullard commented publicly, "I expect most jewelers and gem dealers will be stunned. It seems quite possible that this bill, if it is passed, could touch off similar legislation in other states. The fact that this calls for a civil penalty is the key. It allows the jewelers' own customers to be the enforcing body. Without a civil penalty, jewelers would remain so far removed from the enforcement agency that they would be effectively shielded from the law. The civil penalty imposes a clear and immediate liability on any jeweler who does not comply."

Introduction of the Pennsylvania bill, with all the attendant publicity, seemed to be a legislative endorsement of untreated gems, and indirectly, of the Yogo sapphire. In April, 1985, Intergem's fortunes appeared to be at an all-time high. A record number of jewelers and gem professionals had booked summer tours of the Yogo dike, and NBC's *PM Magazine* was putting together a segment on Intergem and the Yogo sapphire. Market expansion was proceeding on schedule, helped along by growing consumer acceptance of the Yogo sapphire and continuing favorable exposure in the national media. But beneath it all were some serious and growing problems.

When Intergem received the bulk of its corporate financing from Citibank, the value of gold, diamonds and other precious gems had just peaked at historic highs. The primary collateral for the bank loans had been Intergem's working inventory, which included everything from rough stones to cut gems to finished jewelry. In only four years, the value of gold and gems on world markets had declined dramatically, significantly reducing the real value of Intergem's collateral.

The rapid expansion of the market was a capital-intensive operation with high costs in advertising, public and retailer relations, and maintaining a large staff. Sales figures, while very impressive on paper, were tempered somewhat by generous dealer incentives and liberal return-exchange arrangements, thus cutting back on cash flow.

Intergem was burdened by two major financial obligations-payments on the mine property and the interest on the bank loan. In

1981, Intergem had agreed to purchase the mine property from Sapphire International Corporation, now Roncor, Inc., of Los Angeles, California. Intergem paid $1.5 million down and agreed to pay off the remainder of the debt in 14 semi-annual payments of approximately $250,000 each. Meanwhile, interest expense on the bank loan had increased continuously, over 200 percent in the last year alone. Through it all, Intergem had yet to record its first annual profit, although, encouragingly, monthly profits had been realized in September, October and November, 1984. The first sign of a serious cash flow problem came in May, 1985, when Intergem failed to meet a $250,000 payment to Roncor, Inc.

It still seemed that Intergem could realize its first annual profit with the expected record sales of the 1985 Christmas season. But, in June, Citibank indicated that Intergem's line of credit was in danger. President Dennis Brown resigned immediately to "pursue other interests." The first portent of disaster came shortly thereafter when Citibank, concerned with the declining collateral and inventory values and cash flow problems, called in the loan with both interest and principal payable in thirty days.

If Intergem had made a mistake, it was in assuming that its line of credit would be continuously extended, in keeping with its market expansion and increasing sales figures. The timing for the loan recall couldn't have been worse. Over $1 million in orders were on the books for the coming Christmas season and more were coming in every day. But without operating capital, and Intergem now had virtually none, the jewelry to fill those orders would never be manufactured.

During the summer of 1985, as trade journals hinted, some with apparent relish, at Intergem's growing difficulties, the company sharply reduced both staff and operating expenses. To generate cash flow, $78,000 worth of jewelry in discontinued styles was liquidated at prices slightly below cost. In September, Intergem announced a major "reorganization," which included a manufacturing agreement with another company to hopefully fill at least some orders while avoiding borrowing to meet inventory and manufacturing costs. Harry Bullock, in addition to his position as board chairman, took over the duties of president and CEO; Steven Droullard joined the board and accepted the position of senior vice-president.

The trade journals no longer merely hinted at Intergem's troubles. As the company tried to buy time from its creditors, the November,

1985 *Jewelers' Circular/Keystone* featured a cover photograph of Intergem's sapphire recovery plant at Yogo. The cover headline announced a feature article about Intergem.

**INTERGEM AT THE BRINK:
CAN YOGO SAPPHIRE
FIRM SURVIVE?**

The flamboyant minermarketer of Montana's deep blue Yogo sapphires has courted media attention and stirred up the jewelry industry by claiming it is one of the few suppliers of untreated gems. But now Intergem is in deep financial trouble that even threatens ownership of its mine.

The investigative article went into considerable detail about Intergem's financial woes. While painting a bleak picture for the future, it also included statements from Intergem executives. Regarding the Citibank and Roncor obligations, Droullard stated, "Intergem will come up with agreements and settlements in anywhere from thirty days to six months. Since we are negotiating, it is already a foregone conclusion that they have agreed to wait."

Harry Bullock offered this prediction: "You can look at the (financial statements) all you want, but they won't look good for awhile . . . I'm telling you Intergem is going to survive."

By November, 1985, Intergem had fallen $500,000 behind on the Roncor payments and faced increasing pressure from its big creditors. It was also beleaguered on yet another front. The mining picture at the Yogo dike had become clouded through changes in the business and political climate of Montana. When Intergem began work on the Yogo dike five years earlier, it was responsible to five Montana state regulatory agencies. By 1985, that number had increased to 21 and covered everything from land reclamation to water quality to transportation.

The biggest concern, however, was a pending sharp increase in Montana property and mining taxes. In Helena, Harry Bullock argued for fairness in mine taxation by explaining the mechanics of sapphire mine production. It was not fair, Bullock declared, to impose a tax on gross mine production, that is, on every sapphire taken out of the ground, since only about 40 percent of the sapphires mined were of gem quality and thus had material value. Considering the projected $2 million start-up costs of the underground mining operation, Bullock believed the proposed tax increase was economically unrealistic. His arguments, however, did nothing to stop legislators' demands for higher taxes which promised only to dramatically increase costs of future mining at the Yogo dike.

The combination of cash flow difficulties, overdue financial obligations, the poor long-range mining outlook, and a disastrous 1985 Christmas sales season meant the inevitable decision regarding the future of Intergem, Inc., could not be put off much longer. As Bullock and Droullard struggled with the decision, Pennsylvania State Senate hearings on Bill 827 began in the State Capitol in Harrisburg, offering Intergem a chance to fire its parting shot.

Testifying in support of mandatory gem treatment disclosure before the Pennsylvania State Senate Committee on Consumer Protection and Professional Licensure, Droullard stated, "Jewelers do not disclose because of a conspiracy—they were just hoping (the question of gem treatments) would never come up. They have no incentive to disclose without a law. Their attitude is: Why rock the boat? The consumer doesn't need to know. But the time to rock the boat has come.

"This bill admits there is a need for some external enforcement, and thus brings distaste. What is done in Pennsylvania will have an effect around the world. Since the threat of state action, there has already been a flurry of activity in the industry to disclose. And as to the argument that the average jeweler doesn't know if a gemstone has been treated, *somebody* knows in every case—the person who has done the treatment."

Intergem, for five years, had rocked the jewelry industry's boat, but now the rocking was over, along with Intergem's tenure at the Yogo dike. A settlement of the bank debt, both principal and interest, was reached by handing over the working inventory, which included everything from rough sapphires and cut gems to finished jewelry. In summer, 1986, the Yogo mine property, the heart of Intergem, reverted back to Roncor, Inc., thus freeing Intergem of all debt. Intergem then merged with a California-based financial company, leaving its stockholders unscathed, but also, at least for the time being, out of the gem and jewelry business.

The causes of Intergem's demise, which elicited few tears from the jewelry industry, were in some ways similar to those which brought about the closing of the English Mine 60 years earlier. Both had encountered increased production costs and taxation during a period of general weakening of world gem markets. Some observers in the jewelry industry, however, simply felt Intergem had tried to go too far too fast, and had stumbled in the process.

For the first time since American ownership and operation of the Yogo dike resumed in 1956, a company had effectively put together all the diverse aspects of a vertically integrated gem operation, from mining the rough stones through cutting and jewelry manufacture to promotion and marketing. Intergem had brought the Yogo sapphire back to a prominence unequalled since the days of the English Mine. It had certainly made the American jewelry industry sit up and take notice of the Yogo sapphire, while initiating new ideas and standards that promised to have long-term effects on gem and jewelry marketing ethics. Intergem's chapter at the Yogo dike had been dramatic, exciting and highly significant, yet, in the end, it would be remembered as another effort that had come up short.

The passing of Intergem left the future of the Yogo dike, at least in terms of mass mining and marketing of sapphires, unclear. The 1,860-acre mine property had reverted to Roncor, Inc., which immediately put the property up for sale. Any buyer contemplating a mass mining and marketing operation, however, will face a costly reconstruction of the cutting and marketing network. Another consideration for potential large-scale miners is the need for geological data before commencing mining operations. The computerized, in-depth data obtained and refined by consulting geological engineer Delmer Brown from years of surveys, which would take several million dollars to duplicate, are still the property of Intergem.

Still, the Yogo dike is an enormously valuable piece of property. Its true value may be realized only when compared with the world's other six commercial sources of gem sapphires—the Kashmir, Burma, Cambodia, Thailand, Australia and Sri Lanka.

The fabled sapphires of the Kashmir—which Yogos were once represented as—are available only in very limited quantities and at exorbitant prices. Kashmir mine production has always been sporadic and unreliable; even access to the deposits is hindered by extremes of isolation, terrain, elevation and climate. Few professional geologists have ever visited the site and commercial mining is not expected to resume. Since 1974, the Kashmir state has been politically troubled by China's occupation of nearby Tibet, then by the 1965 and 1971 Indo-Pakistan wars, next by the 1979 Soviet invasion of Afghanistan, and most recently by growing conflict with its mother country, India. Nearly 20 years have passed since the last public sale of rough Kashmir sapphires. Although tribesmen still recover a few stones from the police-guarded mines, Kashmir sapphires are no longer a factor in the gem markets.

The Indo-Chinese peninsula may be even more politically unstable and inaccessible. Little information regarding Burma's famed Mogok Stone Tract has reached the Free World since 1962, when a socialist regime banned foreigners and nationalized all industry, including gem mining, to dramatically curtail the supply of Burmese sapphires and rubies. A few stones are still sold legally at annual government auctions in Rangoon; others are illegally smuggled across the Thai border along a route notorious for guerilla warfare, extortion by both Burmese Army and guerilla chiefs, and rampant opium smuggling. opium smuggling.

With the effective closure of Burmese gem sources, Thai (Siamese) mines have recently assumed a growing importance. In 1980, over 20,000 individual miners in the Chanthaburi-Trat area, the primary gem mining district, took out nearly 30 million carats of rough ruby and sapphire. More sophisticated methods of heat treatment are increasing the number of marketable Thai gems. The Thai mines are very near the border of Cambodia, a nation which, since 1979, has been occupied by the Vietnamese Army. The rich Cambodian gem mines at Pailin are today located within sight of a Vietnamese Army headquarters. Few gemstones are believed to have come from Cambodia in recent years.

Currently, Australia and Sri Lanka provide the bulk of the world's sapphires. While Australia enjoys political stability, its blue sapphires are often exceedingly dark; even with heat treating to lighten their color intensity, they may appear lifeless. Some Australian sapphire miners are concerned that their alluvial deposits may be exhausted as early as the year 2000. Sri Lanka's extensive alluvial gem fields have produced steadily for centuries. Historically, mining methods have been primitive, but a gradual move toward mechanization in the government-controlled gem fields indicates future production increases are likely. Sri Lankan sapphires occur in a wide variety of colors; virtually all blues are heat treated to enhance their appearance.

All foreign sapphire sources, with the exception of a small Kashmir deposit, are alluvial, or secondary, deposits. Few geological surveys have ever been performed to indicate their extent and thus their potential for future production. The gem fields of Sri Lanka and Burma have been mined since antiquity, and those of Thailand, Cambodia and Australia for a century. Historic production records do not exist; even today, especially in the oriental gem fields, rampant smuggling and illegal mining make estimations of current production uncertain.

The Yogo dike, however, stands apart from every other known gem sapphire source. It enjoys the highest degree of political stability; geographically, its location limits mining only on a temporary, seasonal basis during the Montana winters. Geologically, the dike has enormous proven sapphire reserves; gemologically, its sapphires are among the finest in the world and the only such stones that can be certified *en masse* as natural. Not surprisingly, many consider Montana's Yogo dike as the world's premier gem sapphire deposit.

Even before the Intergem era had passed, a group of local Utica residents were writing a new chapter in the story of the Yogo dike. Like many men whose family roots went back three generations into Judith Basin country history, Lanny Perry and Chuck Ridgeway had not only an interest in Yogo sapphires, but practical knowledge of mining to back it up. Both had prospected, sluiced gold and performed assessment work on gold mines in the Little Belt Mountains, mostly in the area of Yogo Gulch. Lanny Perry had even worked for Sapphire International Corporation during the early 1970s when the "Kunisaki Tunnel" was being developed at the site of the American Mine on Yogo Creek.

In January, 1984, Lanny Perry and Chuck Ridgeway, along with their wives, Joy and Marie, set out on a day of prospecting in an area that had never been mined—the rugged hills between Yogo Creek and Kelly Coulee, where the westernmost known extention of the Yogo Dike terminated. What followed was a discovery that would have made Jake Hoover or Jim Ettien proud. Making his way down a steep slope in difficult terrain, Ridgeway began sliding and dislodging loose dirt and rocks. Lanny Perry stopped in his tracks and carefully examined the loosened alluvial material. It looked exactly like the sapphire ore he had helped mine at the Kunisaki Tunnel a decade earlier. Filling their day packs with the crumbling rock, the group hiked out. That evening, they screened the dirt and recovered a single rough Yogo sapphire. The Perrys and Ridgeways had discovered a previously unknown section of the Yogo dike.

With exploration work limited to weekends, the group packed in picks and shovels to slowly and laboriously determine the extent of their discovery. By the end of February, 1984, they had staked and recorded the "Sure Thing," "You Bet," "Shoestring," and "Fair Weather" claims, fourteen in all. Although much geological investigation remains to be done, the claims apparently cover a section of the Yogo dike which may have been displaced by local faulting a short distance north of the dike's primary east-west axis. During the following year, they determined that their discovery had the typical geological configuration of the remainder of the Yogo dike—a thin, linear section of dike rock contained between two walls of ancient limestone. Even with exploration limited to pick and shovel work, and mining limited to hand screening, washing and sorting, the Perrys and Ridgeways were accumulating a small, but growing collection of rough Yogo sapphires.

By spring, 1985, they were faceting the larger stones themselves and beginning to sell them locally. Sensing a small, but potentially profitable market, the Perrys and Ridgeways decided to go commercial. In June, 1985, they formed Vortex Mining, headquartered in Utica. The little company was soon joined by Pete Ecker, a longtime friend and former Sapphire International Corporation employee, who provided practical engineering experience. The last partner to join Vortex Mining was Paul Davis, Jr., another experienced local miner who held a number of placer gold claims on upper Yogo Creek.

When Vortex Mining was formed, the Yogo sapphire was enjoying an unprecedented wave of publicity thanks to the national advertising and marketing efforts of Intergem, which was then at its peak.

While the growing demand for Yogo sapphires benefitted Vortex, the little company was clearly in the shadow of a giant. "Some jewelers were hesitant to order from us," Chuck Ridgeway said. "They thought we were some kind of fly-by-night operation. Everyone had heard of Intergem, but we were just too small for some people to take seriously."

Still, Vortex continued working, expanding their mining operation with trucks and light mechanical equipment. Intergem granted access across its own property, making the Vortex claims more easily accessible and mining more efficient and economical. Vortex also developed a system of vertical integration, similar to that employed by Intergem, however on a much smaller scale. The Vortex partners performed all mining, sapphire recovery and sorting. Rough stones over .35 carats were faceted and polished by Lanny Perry and Chuck Ridgeway; the smaller stones were consigned to an agent in Seattle, then forwarded to Bangkok, Thailand, for cutting.

Vortex's marketing is still largely a matter of word of mouth, but has received a strong boost by the demise of Intergem. Individual buyers and a growing number of Montana jewelers are now turning to Vortex Mining as a reliable source of Yogo sapphires. Although most sales are of loose, cut gems, Vortex is looking into the manufacture of finished jewelry for the future.

There are many people in the gem trade who don't consider tiny Vortex to be a worthy successor to the multi-million-dollar corporate approach employed by Intergem. But Chuck Ridgeway offers an interesting point of perspective. "Sure Vortex is small," he admits. "And, at least for the short term, we'll probably stay small. We're trying to match mining and marketing while offering a quality product. Just because something is big doesn't mean it's better.

"We've watched a lot of people come to the Yogo dike over the last twenty years," Ridgeway continues. "Most of them were big, high-profile operations that had a lot of money to throw around. Well, we're here working and they aren't."

Chapter 8

A Century of Sapphires

The decade of the 1990s, which marked a century in the saga of the Yogo sapphire, began on a quiet note. After the abrupt end of the "Intergem era" in 1985, Yogo sapphires had faded from public attention. Nevertheless, small-scale mining operations continued at the Yogo dike. Vortex Mining pursued exploration and development work on its recently discovered dike extension, while Roncor, Inc., owner of the main section of the dike, also resumed mining and marketing. But in 1991, in testimony, perhaps, to the eternal lure of Yogo sapphires, activity picked up once again at the Yogo dike.

Vortex Mining, holding 300 acres of unpatented mining claims along the dike's western extension, worked its claims on the southwest side of Yogo Creek. In June, 1991, a graduate geology student from the University of Calgary visited to research a thesis on the dike extension, concluding that it was not as geologically similar to the main dike as previously thought. He identified maar surge ring deposits, indicating origin not as a magmatic fault emplacement, but as a diatreme blow-out pipe. Core drilling to 110 feet soon defined the wall of the diatreme throat and intersected pockets of sapphire-rich dike rock.

By 1992, Vortex miners had sunk a 70-foot-deep shaft and driven 300 feet of exploratory drifts on the 60-foot level. Vortex initially hand-trammed ore to the shaft station, but later mechanized the operation with a small diesel load-haul-dump unit. Miners raised ore to the surface in an eighteen-cubic-foot-capacity, auto-dumping hoist bucket, then chute-loaded it into a truck for haulage to a washing plant 300 yards east of the shaft.

In the course of exploration, Vortex encountered numerous ore-bearing matrix pods over a large area of brecciated rock and recovered thousands of carats of Yogo sapphires. While some pods were small, others contained more than 100 cubic yards of sapphire-bearing matrix rock.

In 1994, when the Vortex shaft reached 200 feet, miners drove a 135-foot-long exploratory tunnel into two veins of sapphire-bearing, diatreme-type rock. The rock is so highly altered that it breaks down after just a few days of exposure to water and air. Miners have explored only the smaller of the two veins, encountering high grades of sapphire-bearing rock. Vortex estimates the second vein may be as wide as 100 feet, which would constitute a significant commercial source of sapphires.

Encouraged by its sapphire discoveries at depth, Vortex plans to sink its shaft as deep as 500 feet and core drill to 1,000 feet. Exploration and production mining in the vein areas proceeds simultaneously, enabling Vortex to continue to build up its sapphire inventory. Vortex markets both rough and gem sapphires, faceting stones over .35 carats in nearby Utica and consigning smaller stones to cutters in Bangkok.

Roncor, Inc., conducted small-scale mining on its main section of the Yogo dike, recovering, cutting, and marketing sapphires. Roncor miners found a remarkable stone in 1992. Reporting the find in early July, the *Great Falls Tribune* also mentioned the long-standing dispute between Roncor and Sapphire Village residents over the issue of "mining rights."

GIANT SAPPHIRE FOUND NEAR UTICA

A sapphire mine near Utica has yielded a blue stone that may be worth more than $100,000. Roncor, Inc., a commercial mining company, unearthed an 11-carat Yogo sapphire in late June. The *Tribune* was told of the find last week. "It's big news for us. . . ," said Joe Iwawaki, Roncor's mining superintendent. Iwawaki says he's sure the stone's value is in the six-figure range. The stone, described as a "perfect specimen of cornflower blue Yogo," is believed to be the largest found at the mine since about 1910. Iwawaki said stones in the five- to seven-carat range are big finds at the mine at Sapphire Village, about ten miles southwest of Utica. . . .The stone has been sent to

Roncor headquarters in California, Iwawaki said. He's not sure if the company will have the sapphire cut for resale or keep it as a specimen. Roncor is processing about 100 tons of dirt per day at the mine, Iwawaki said. The company, which has operated the mine since 1986, is employing seven workers this summer. A volcanic dike bearing sapphires runs for a couple of miles along a ridge west of Sapphire Village. Residents of the village have used buckets and shovels to mine portions of the dike. Some of the residents believe that Roncor has kept them from mining in portions of the dike that hold the most sapphires.

A group of residents sued but the suit was dismissed.

By far the most significant event of the early 1990s at the Yogo dike was the first active involvement of a major American mining company. In the late 1980s, geologist Delmer Brown, who had investigated the Yogo dike during the Intergem years, brought the mining and marketing potential of Yogo sapphires to the attention of Willem Lodder, president of AMAX Exploration, Inc., the exploration arm of AMAX, Inc., a major mineral resource development corporation. Although AMAX traditionally focused on far larger development projects, Lodder believed Yogo was worth investigating.

Together with Cheni S.A., a French mining company, AMAX Exploration organized the Yogo Sapphire Project and began planning a two-year program aimed at comprehensively evaluating the Yogo sapphire resource and ore reserves, determining costs for a 500-ton-per-day underground mining operation, and testing the sapphire market. In March, 1993, AMAX Exploration, the operating partner of the joint venture, entered into a twenty-two-month lease-purchase agreement with Roncor, Inc.

Field operations began on the $2 million project in June, 1993. Under direction of project manager Bob Johnson, an AMAX Exploration mining engineer, crews began work on the main section of the dike at two sites, the "East Site" and the "Middle Mine Site." Miners drove tunnels declined at twenty degrees to intersect the dike at points about 50 feet below the floor of the old surface trenches. They then drove 200 feet horizontally along the dike at the East Site, and 100 feet along the dike at the Middle Mine Site. Using a delayed cut-and-fill, or "backstoping," mining method, they extracted a total of 7,500 tons of dike rock.

Underground mining proved difficult, particularly at the East Site, where the adjacent country rock was not the expected solid limestone, but a soft, unstable formation of sandstone and shale that required extensive ground support. Central Montana's coldest, wettest summer on record created another problem reminiscent of the cloudburst that had halted operations at the English Mine seventy years earlier. In late August, 1993, torrential rain followed by six inches of snow produced a heavy runoff that flooded the Middle Mine Site workings.

The degree of dike rock alteration varied widely between the two sites. Rock from the Middle Mine Site decomposed very quickly upon exposure to air and water, while rock from the East Site took considerably longer. After decomposition, crews washed the dike rock in a trommel, then hand-sorted the sapphire concentrate. Gem-quality rough was sent to Bangkok for cutting.

Exploration furthered the idea that the Yogo dike is but a single feature in what may be a complex system of nearby dikes. Geologists and sapphire miners had long known of the existence of a sapphire-barren dike to the north of, and running parallel to, the main Yogo dike. And in the early 1980s, Vortex had discovered a previously unknown, sapphire-bearing, northwest "extension" of the main dike.

In 1994, AMAX Exploration added to the known system of nearby dikes with yet another discovery. Geophysicists performing a magnetometer survey noted a magnetic anomaly northeast of the main dike. The anomaly indicated a narrow, straight-line feature about 500 feet northeast of and parallel to the main dike. Yet no slight trench-like subsidence or other surface feature was present to visually indicate the presence of another dike.

Nevertheless, when crews dug through six to fifteen feet of overburden, they exposed the in situ, sapphire-bearing rock of what is now known as the "Eastern Flats Dike." To test sapphire grade, crews trenched the new dike, which averaged about four and one-half feet in width, for a length of 300 feet, removing some 1,000 tons of dike rock. Yogo sapphires were present, but in relatively low grades.

Trenching provided another interesting discovery—the remains of an old, filled-in, twelve-foot-deep shaft that had been dug to reach the in situ dike rock. Whoever dug the shaft, probably during the years of the English Mine, apparently sampled the dike rock, but found it to be too low-grade to mine. Judging from the animal bones and broken china found in the fill material, the shaft was used as a refuse pit after it was abandoned. The lingering question is, how did the unknown miner or miners seem to know exactly where to dig through deep overburden to reach the dike when there was no surface indication of its existence?

By the end of 1994, the Yogo Sapphire Project had collected enough data to draw three conclusions regarding the Yogo dike, the marketability of Yogo sapphires, and the overall costs involved in a commercial-sized mining venture.

First, the project had set its size "cut-off" limit for rough sapphire recovery at three millimeters. While some stones cut into magnificent gems as large as 1.8 carats, 80 percent of the rough sapphires were far smaller stones ranging between three and five millimeters. These small stones yielded cut gems averaging less than ten points (one-tenth of a carat) in weight, a size suited only for melee-type jewelry use.

Second, nearly a century of off-and-on mining had virtually depleted the easily accessible, near-surface dike rock. Accessing any significant volume of sapphire-bearing dike rock in the future would therefore require underground mining. Because of the high estimated cost of a 500-ton-per-day underground mining operation, the price of Yogo sapphires would be non-competitive with the less expensive foreign sapphires now available.

Third, although the overall sapphire resource—that is, the total amount of sapphire present within the dike rock—is enormous, the Yogo dike actually has very limited proven reserves of ore, meaning sapphire-bearing dike rock that can be mined at a profit.

After considering the approximate $10 million cost to purchase the Yogo dike property, the very substantial additional capital necessary to secure state permits and environmental bonds for a 500-ton-per-day commercial mine, the high cost of underground mining, and the small general size of the sapphires, the Yogo Sapphire Project joint venture elected to not exercise the purchase option of its contract.

As AMAX Exploration retired from the Yogo dike in 1995, another significant supply of Yogo sapphires appeared on the market from an unexpected source—Intergem. Before its 1985 failure, Intergem had aggressively promoted Yogo sapphires in many areas of the United States where jewelers typically had never heard of the stones, much less carried them in stock. But after greatly furthering national awareness and acceptance of Yogo sapphires, Intergem's demise created two problems: the national promotional effort stopped and the supply of Yogos became erratic.

As Intergem's cash flow worsened during its last months, the company paid its sales representatives not in cash, but in sapphires. After Intergem collapsed, many of its former salespeople continued to make their rounds, selling their own sapphires until their supplies were gone. As market supply became tighter and more erratic, many jewelers, including those who had done well with Yogos in the past, liquidated their remaining stocks and went back to selling Australian and Oriental sapphires.

When Intergem collapsed, Citibank, one of the company's prime creditors, ended up with a huge stock of Yogo sapphires, including 200,000 carats of rough, 22,000 carats of cut gems, and some 2,000 pieces of finished jewelry. That entire inventory was thought to be worth perhaps $3.5 million. But with no in-house expertise in gemology or gem and jewelry marketing, Citibank was unable to sell the sapphires or to even accurately estimate their worth. Rather than attempting to sell the "Intergem legacy" in piecemeal lots, Citibank first tried to dispose of the sapphires and sapphire jewelry in a convenient single sale. But when a New York City gem and jewelry wholesaler placed the top bid at only $275,000 for the lot, Citibank rejected the offer.

The Intergem sapphires then gathered dust in a Citibank vault until 1992, when Sofus Michelson, director of the Center for Gemstone Evaluation and publisher of the Michelson Gemstone Index, offered to help liquidate the stones. Michelson sought the expert advice of jeweler Jim Adair, whose Adair Jewelers of Missoula, Montana, was among the world's leading retailers of Yogo sapphires.

When Adair arrived in New York City, armed guards transported the four sealed canvas bags of Intergem sapphires, which had been stored for six years in a nearby underground vault, to the Citibank offices. As bank officers watched, Adair removed the seal on the first canvas bag, reached inside—and withdrew a strange material accompanied by a cloud of dust. As bank officers coughed, Adair studied the material in his hand.

"Well?" a Citibank officer asked impatiently. "What is that stuff?"

The drama of the moment notwithstanding, Adair couldn't help but laugh. "It's dirt—ordinary dirt!" the Montana jeweler announced.

During Intergem's last days, someone, either in an attempt at black humor or to deceive Citibank as to the amount of sapphires it was actually receiving, had apparently filled a canvas sack with dirt and a few worthless sapphire chips, sealed it, and passed it on to the bank as another sack of "Yogo sapphires."

The contents of the second bag that Adair inspected were a little better, with some gem-quality stones mixed in with reject material. The third bag was better yet, but still not good enough to satisfy the Citibank officers. The bank officers finally breathed a collective sigh of relief only when the last bag was opened to reveal a fortune in fine rough and cut Yogo sapphires.

Adair purchased the finished jewelry and a selection of rough and cut stones from Citibank. Rather than cutting the roughs into traditional standard gem shapes, he and Michelson designed entirely new shapes to better display the brilliant life and delightful cornflower blue color inherent to Yogo sapphires.

Vortex Mining, based in nearby Hobson, Montana, continues as the only source of newly mined Yogo sapphires. In 1984, when the original members of Vortex Mining had discovered and claimed a previously unknown extension of the Yogo dike, their accomplishment had been overshadowed by Intergem's highly publicized mining and marketing activities. But today, thanks to a balanced development program that includes thorough exploration, good mining practices, conservative financial planning, and a solid marketing plan, Vortex Mining operates the sole producing mine on the Yogo dike.

In 1997, Vortex prepared for full production by erecting a new headframe and installing an electric hoist and a new generator system. The shaft is now 265 feet deep, with shaft stations located at intervals below the 60-foot level. In 1999, at the 180-foot level, Vortex miners found two rich, sapphire-bearing veins of dike rock 120 feet apart. They are now mining the vein closest to the shaft station and will eventually drive a crosscut to intersect and mine the far vein.

At the 180-foot depth, ore grades contain an average of about 30 carats of gem-quality sapphires per ton. With annual ore production currently exceeding 1,500 tons, the mine now provides more than 45,000 carats of cuttable, rough Yogo sapphires each year—stones that are considered to be the highest-quality sapphires ever taken from the Yogo dike. Recently, one big rough sapphire cut into a spectacular 3.03-carat, round, brilliant gem valued at $50,000. Vortex has also mined another gem-quality rough with a weight of 10.61

carats. That stone will not be cut but will instead be sold as a specimen representing the very best of the Yogo dike

Along with providing a steady supply of new Yogo sapphires, Vortex Mining is also contributing to a better understanding of the geology of the Yogo dike. With the Vortex shaft permitting visual inspection of the Yogo dike to an unprecedented depth of 265 feet, the dike's true geological complexity has become much more apparent. Because of significant variations in sapphire concentration, Vortex Mining has proven that profitable mining on the Yogo dike is not a simple matter of digging and processing dike rock, but rather one of thorough exploration and precise determination of production mining sites in terms of both depth and lateral location.

After fifteen years of development on the Yogo Dike, Vortex Mining is just now hitting its stride. In late 2001, Vortex will complete development of a 2,500-foot-long declined tunnel that will access the 400-foot-level of the Yogo dike to mine rich ore bodies that core drilling has already delineated. The new Vortex decline will triple production of Yogo sapphires by mid-2002.

Lanny Perry, one of the original members of Vortex Mining, sums up his company's approach to mining the Yogo dike. "We take it one step at a time," Perry says. "We're not trying to get rich quick. Those who tried to are long since gone. But we're more encouraged now than ever before. Hopefully, in the long run, we'll be able to say that we brought America's Yogo sapphire back to prominence as the world's premier sapphire."

Today's growing interest in the Yogo sapphire reflects nothing more than the rediscovery of the stone's fine quality and great beauty, attributes that had been recognized from the beginning. After earning awards at the turn-of-the-century Paris and St. Louis expositions, Yogo sapphires went on to gather many other honors. By 1918, Yogo sapphires appeared in the personal gem collections of the Duchess of York, Princess Mary and Queen Victoria of England, and Kaiser Wilhelm of Germany. Historians also believe that Yogo sapphires, possibly misrepresented as "orientals," were acquired for the British Royal Crown Jewel Collection. In 1920, Johnson, Walker and Tolhurst, Ltd., presented four cut Yogos, thirty roughs, and specimens of the dike rock matrix with visible imbedded sapphires to the British Museum of Natural History in London, where they are still exhibited today.

Tiffany & Co., of New York City, one of the earliest champions of the Yogo sapphire, created what is still the most magnificent piece of Yogo sapphire jewelry ever made—a large iris-shaped brooch containing 120 Yogo sapphires. These stones, mined by Charles Gadsden's men in the early 1900s, were sold to Tiffany by the New Mine Sapphire Syndicate. In 1909, the Tiffany Iris Brooch, as the famous piece was known, was purchased by Henry Walters, an avid jewelry collector who patronized contemporary artists. The brooch, with its 120 Yogo sapphires, has recently completed a ten-city, three-year traveling exhibition called "Objects of Adornment: Five Thousand Years of Jewelry from the Walters Art Gallery, Baltimore," a tour sponsored in part by the National Endowment for the Arts.

One of the earliest American collectors of Yogo sapphires, of course, was Dr. George Frederick Kunz, who often acquired fine stones for the renowned gem connoisseur J. Pierpont Morgan. The Morgan stones, which include twenty-nine Yogo sapphires, two of which are three-carat gems, are today part of the gem collection of the American Museum of Natural History in New York City. Another major American collection is that of F. G. McIntosh, assembled in the 1920s. The McIntosh stones, which include eighty-three Yogo sapphires, are now exhibited with the gem collection of the California Institute of Technology in Pasadena. Yogo sapphires are also part of the gem collection of the United States National Museum (Smithsonian Institution) in Washington, D.C. This collection boasts a 10.2-carat stone which is believed to be the largest cut Yogo in existence.

Yogo sapphires have also served as gifts to United States Presidents. In 1923, a Great Falls jeweler mounted a Yogo sapphire in a ring made of Montana gold, presenting the "all from Montana" gift to Mrs. Warren G. Harding. And in 1952, in his last effort to promote his beloved Yogo sapphires, Charles Gadsden presented cut Yogos to President and Mrs. Harry S. Truman and their daughter, Margaret.

That list of honors is quite impressive for a sapphire that was, at various times, too small, too flat, too expensive to mine, misrepresented and lacking in image. Yet, today, the Yogo sapphire is even rising above its old honors and awards; with its origin proudly proclaimed, the Yogo has earned market acceptance in direct competition with foreign sapphires. The name and image of the Montana sapphire is not likely to suffer any further before the grand names of the oriental stones. Orientals may be large, but few match the color and beauty of the Yogo. And in that controversial matter of heat treatment, no sapphires in the world can challenge the quality of the Montana sapphires that come from the Yogo dike. The standard of excellence among sapphires, finally, is right back where it belongs—Montana and the Yogo dike.

Even today, the Yogo dike, and the country that surrounds it, is part of a different world, that of Jake Hoover, Charlie Russell and Charles Gadsden. A visit here is a long step back in time. Most visitors come from Lewistown, driving west on U.S. Route 87 across the farm and ranch lands of the sprawling Judith River Basin. They cross the Judith River, then turn off the highway at the tiny town of Hobson, named for Simeon S. Hobson, the rancher and, later, Montana state senator who grubstaked a local hunter and prospector named Jake Hoover in his search for gold in the nearby Little Belt Mountains nearly a century ago. From Hobson, an asphalt county road follows the Judith River to Utica, the little cow town that was painted into history by Charles M. Russell. Nestled at the edge of the Judith River Basin and the Little Belts, Utica is a quiet place that time has somehow passed by. The main street is still only two blocks long, just as it was when Russell painted his famed watercolor *A Quiet Day in Utica*. Simple, low, old buildings border the wide, dusty streets, silently guarding the many memories of yesterday, memories that seem hauntingly real. This is the same street where Hoover, Hobson, Russell and Millie Ringgold once walked and where the stagecoaches once rolled, stagecoaches that, many years ago, brought Englishmen in and took canvas bags of Yogo sapphires out.

A gravel road continues to the southwest along the Judith River, gradually climbing into the foothills. Eleven miles from Utica and somewhat incongruous with the pleasant, open country, stands a cluster of mobile homes and campers on the slopes west of the river—Sapphire Village, where lot owners still hold the right in perpetuity to dig nearby for gemstones. A half mile upstream is the confluence of the South Fork of the Judith, a place still known as Pig-Eye Basin. It was here, a century ago, that Jake Hoover built his tiny cabins and, for many years thereafter, offered a meal and a place to sleep to any weary rider who passed his way.

A smaller dirt road turns west and climbs into the hills where, a half mile farther, it is blocked by a locked gate. Only one hundred yards beyond that gate is what first appears to be a long trench—the eastern end of the Yogo dike. For nearly five miles, the dike slashes arrow-straight across the earth like a great scar. In some places, it is still only a faint depression, just as Jim Ettien first saw it; in others, it has been excavated to a depth of twenty-five feet, exposing the smooth walls of the Madison limestone that was laid down by an ancient sea 300 million years ago. Near the middle of the dike, one can still find the ruins of the English Mine. The concrete holding dams are nearly buried by the gravel that swept over them in the great cloud burst of July 26, 1923. Foundations of long gone mine buildings lay hidden in the waving wheat grass and occasional scattered timbers remind one of the once elaborate systems of sluices, flumes, tramways and weathering floors that once stood here. And in dark, mysterious caverns within the dike, the timbers and stulls that still support forgotten workings are visible. Remarkably preserved, the Gadsden House stands like a lonely sentinel atop the low hill overlooking the middle of the dike. The windows where Charles Gadsden once stood are boarded; they stare like empty eyes over the ruins of the mine that produced $25 million in cut sapphires for the British.

The land is still remote and isolated; with the exception of the workings that scar much of the surface of the dike, it is as beautiful as it was when Indians hunted here. To the south is the vast Lewis and Clark National Forest; to the north are the pine-covered hills of the Judith River Game Range, a state game management area that is home for large, roaming herds of elk and deer, descendants of the animals that fell before Jake Hoover's thundering Winchesters long ago. A bit farther to the west, small flocks of pigeons are often seen wheeling over the pines, birds whose ancestors once carried messages to the resident supervisor of the English Mine.

The road now becomes narrow and rutted as it descends Tollgate Hill and passes the site of the old tollgate, the narrow cut blasted out of the limestone in 1879. The armed toll collectors are gone, of course, but set into the limestone high above the road, and often unnoticed, is a small bronze plaque dedicated to the memory of Charles Gadsden. Beyond the tollgate is a clear mountain stream flowing through a forested gulch in the shadow of sheer limestone cliffs—Yogo Creek. The right fork leads upstream; in a few miles it passes the overgrown, rotting ruins of old Yogo City, where the gold discovery that opened the Little Belts was made in 1879. Travel over the left fork of the road is prohibited, but visitors who are granted permission to enter soon find themselves at the point where Yogo Creek has transected and cut deeply into the Yogo dike. Across the creek to the west is Kelly Coulee, where John Burke and Pat Sweeney staked their "Fourth of July" claims in 1896. Not far to the northwest are the active claims of Vortex Mining, the current miner and supplier of Yogo sapphires. On Yogo Creek, all that remains of the American Mine, later the Kunisaki Tunnel, is the locked portal. This was also the site of the Intergem washing plant, and still standing are the cluster of maintenance sheds alongside the weathering heaps and tailings ponds.

A short distance downstream is another site of great relevance to the Yogo story, one devoid of activity and not marked by plaques or ruins. It was here that the greatest precious gemstone discovery in North America was made. In the whisper of the pines and the shadow of the towering limestone walls, Jake Hoover set up his sluice boxes here in 1895. Old Jake never found much gold, but he started America's greatest adventure in mining native precious gemstones when he filled a cigar box with those blue pebbles.

Sources, Acknowledgments and Thanks

The people and sources who were of assistance to me in writing the story of the Yogo sapphire are too numerous to be mentioned in their entirety. Among the most useful published sources were the United States Geological Survey's *Mineral Resources, Annual Reports* and *The Mineral Industry* series. Of special importance was the United States Geological Survey *20th Annual Report* (1900), *Geology of the Little Belt Mountains, Part 3*, by W. H. Weed, and United States Geological Survey Bulletin 983, *Corundum Deposits of Montana* (1952), by Stephen E. Clabaugh. Other books and periodicals that shed light on sapphires and the early years at Yogo were Dr. George Frederick Kunz's *Gems and Precious Stones of North America* and *The Curious Lore of Precious Stones; Precious Stones*, Vol. II, (1904), by Max Bauer; *The Occult and Curative Powers of Precious Stones* (1907), by William T. Fermie, M.D.; *Mining and Scientific Press; American Journal of Science; Lapidary Journal; The Saturday Evening Post; Gems and Minerals* and the United States Bureau of Mines *Minerals Yearbook* series.

Of great help in researching the ties between Charles M. Russell and Jake Hoover were Mr. Jerry Goroski, Assistant Director, C. M. Russell Museum, Great Falls, Montana; Mr. William H. Lang, Editor of the Montana Historical Society's *Montana, The Magazine of Western History*; and the staffs of the Amon Carter Museum and the Sid Richardson Collection of Western Art, both of Fort Worth, Texas.

Many of the Hoover-Russell incidents were excerpted from Frank Bird Linderman's *Recollections of Charley Russell.*

I have never received greater interest, cooperation and assistance in researching any book than I have from the Montanans at the many Montana libraries, historical societies and other sources that all provided parts of the Yogo story. Special thanks to Ms. Florence Kettering, Director, Lewistown City Library, Lewistown; Ms. Barbara Twiford, Utica Museum and Historical Society, Utica; Ms. Mary Ann Quirling, Lewistown Genealogy Society, Lewistown; Mr. Harley W. Yeager, Information Officer, Montana Department of Fish, Wildlife and Parks, Great Falls; the *Great Falls Tribune*; the *Billings Gazette*; Mr. Ken Byerly, Publisher, the *Lewistown News-Argus*; Sister Marita Bartholome, Reference Librarian, Great Falls Public Library, Great Falls; Ms. Nancy Dahy, Reference Librarian, Montana College of Mineral Sciences and Technology, Butte; Mr. Lester G. Zeihen, Adjunct Curator, Mineral Museum, Montana Bureau of Mines and Geology, Butte; Mr. Dave Walter, Reference Librarian, Montana Historical Society, Helena; Ms. Julie Davies, Publicity Coordinator, Travel Montana, Helena; and the staff of the Maureen and Mike Mansfield Library, University of Montana, Missoula.

Of great assistance in compiling data on the modern gem and jewelry trade, as well as the current controversial issue of gem treatment, was the objective and timely coverage provided by *Modern Jeweler, Southern Jeweler, National Jeweler*, and *Jewelers' Circular/Keystone*. Additional information was offered by Mr. Frank Farnsworth, President, Idaho Opal and Gem Corporation, Pocatello, Idaho; and Mr. Roland Naftule, President, American Gem Trade Association, Phoenix, Arizona.

Others who, in their own areas and ways, provided further assistance are Mr. Carl A. Francis, Curator, Mineralogical Museum, Harvard University, Cambridge, Massachusetts; Mr. J. P. Fuller, Assistant Curator, Dept. of Mineralogy, The British Museum, London, England; Dr. H. Stanton Hill, Curator of Minerals, Division of Geological and Planetary Sciences, California Institute of Technology, Pasadena, California; Dr. John M. Shannon, Museum Director, Colorado School of Mines, Golden, Colorado; Mr. Joe Peters, Sr., Scientific Assistant, American Museum of Natural History, New York, New York; W. R. C. Shedenhelm, Editor of *Rock & Gem* magazine, Ventura, California; Mr. John Loring, Senior Vice President and Design Director, Tiffany & Co., New York, New York; Mr. Bill Richards, Staff Reporter, *The Wall Street Journal*; and Mr. Dave Parry, Director, Lake County Public Library, Leadville, Colorado, and his staff.

I'd also like to express my appreciation to two people who encouraged me from the beginning to expand the Yogo story from a group of magazine articles into a book. They are Ms. Michele Wells, Public Relations Consultant and Partner, Wells Communications, Boulder, Colorado; and Mr. Dave Flaccus, Publisher, Mountain Press Publishing Company, Missoula, Montana.

Thanks, too, to all the people of Intergem, Inc., Aurora, Colorado, especially to Mr. Harry C. Bullock, Chairman, who took the time to explain the beginnings of Intergem, and to Mr. Dennis Brown, President, and Mr. Steve Droullard, Vice President/Marketing, who introduced me to the world of modern gem business, marketing and advertising.

I'm also indebted to Marie and Chuck Ridgeway of Vortex Mining, Utica, Montana, for bringing me up-to-date on the most recent developments at Yogo.

My greatest thanks and deepest appreciation must be directed to Delmer L. Brown, consulting geological engineer and gemologist, of Lakewood, Colorado, who is the true authority on matters historical, geological and gemological regarding the Yogo dike and Yogo sapphires. My thanks to Del Brown must cover all of the following: for the days spent instructing me in relevant points of geology and gemology; for the use of his 136-page definitive report, *Geology of the Yogo Sapphire Deposit, Judith Basin County, Montana* (1982); for all the materials from his personal library and files that touched on everything from the history and geology of foreign sapphire deposits and synthetic gems to gem treatments; for the accounts of his personal experiences that provided invaluable insight into the Bangkok cutting industry and the gem fields of Sri Lanka and Thailand; and for all that "hands on" side by side comparison of Yogos with foreign sapphires. For the sake of readability, I've generalized somewhat on a few of Del's geological and gemological ideas and concepts. I fear that Del, as a scientist, may not approve, but ask his forgiveness anyway.

Index